W9-BND-538

TORNADO HUNTER

TORNADO HUNTER

GETTING INSIDE THE MOST VIOLENT STORMS ON EARTH

Stefan Bechtel with Tim Samaras

Foreword by Greg Forbes

NATIONAL GEOGRAPHIC

WASHINGTON, D.C.

Founded in 1888, the National Geographic Society is one of the largest nonprofit scientific and educational organizations in the world. It reaches more than 285 million people worldwide each month through its official journal, *National Geographic,* and its four other magazines; the National Geographic Channel; television documentaries; radio programs; films; books; videos and DVDs; maps; and interactive media. National Geographic has funded more than 8,000 scientific research projects and supports an education program combating geographic illiteracy.

For more information, please call 1-800-NGS LINE (647-5463) or write to the following address:

National Geographic Society
1145 17th Street N.W.
Washington, DC 20036-4688 U.S.A.

www.nationalgeographic.com

For information about special discounts for bulk purchases, please contact
National Geographic Books Special Sales: ngspecsales@ngs.org.

For rights or permissions inquiries, please contact National Geographic Books
Subsidiary Rights: ngbookrights@ngs.org

Published by the National Geographic Society

Library of Congress Cataloging-in-Publication Data

Bechtel, Stefan.
Tornado hunter getting inside the most violent storms on Earth / Stefan Bechtel with Tim Samaras ; foreword by Greg Forbes.
p. cm.
Includes bibliographical references and index.
ISBN 978-1-4262-0302-2 (trade)
1. Tornadoes--United States. 2. Storm chasers--United States--Anecdotes. I. Samaras, Tim. II. Title.
QC955.5.U6B427 2009
551.55'3--dc22

2009000724

Illustrations Credits

Cover: (Background), Gene Rhoden/Weatherpix Stock Images; (insets and spine), Timothy Samaras

Interior: All photographs by Timothy Samaras unless otherwise noted: 19, Carsten Peter; 72, Robert Clark; 129, Rich Thompson/Twistex; 141, Courtesy Dave Hoadley; 148, Carsten Peter; 164, Ted Fujita; 173, AP/Wide World Photos; 187, Ted Fujita; 190, Ted Fujita; 255, Carsten Peter.

Insert: 1 (UP), Carsten Peter; 1 (LO), Carsten Peter; 2 (UP), Carsten Peter; 2 (LO LE), Carsten Peter; 2 (LO RT), Carsten Peter; 3, Carsten Peter; 4-5, Timothy Samaras; 6, Timothy Samaras; 6-7, Jayson Prentice; 8 (UP), Rich Thompson/Twistex; 8 (LO), Jayson Prentice.

Images opening each chapter: Art: drawing adapted from Tetsuya Theodore Fujita. *Memoirs of an Effort to Unlock the Mystery of Severe Storms During the Years, 1942-1992;* Photographic sequences: from video taken by Timothy Samaras in the field

Text Credits: Reprinted with the permission of Simon & Schuster, Inc., from THE MAN WHO RODE THE THUNDER by William Rankin. Copyright © 1960 by Prentice-Hall, Inc. Copyright renewed © 1988 by Prentice Hall, Inc. All rights reserved.

Interior Design: Cameron Zotter and Al Morrow

Printed in U.S.A.

CONTENTS

FOREWORD

THE UNITED STATES HAS MORE TORNADOES THAN ANY OTHER COUNTRY IN THE world. You've seen videos of them and the destruction they cause. What you might not realize is just how many people actually make storm chasing their hobby. Each February a National Storm Chasers Convention is held in Denver, Colorado. I've given talks at several of them, and was amazed that over 200 chasers were in attendance—and that's only a fraction of the number of people who chase storms. Sometimes there may be as many as three dozen chasers following the same thunderstorm! Chasers aren't confined to the Midwest but emanate from most parts of the United States and even from other countries, some making their summer vacation a trip to Tornado Alley. There are even several storm-chase tour companies that take paying customers on the experience of a lifetime.

Chasers love the challenge of predicting where severe storms will develop, then navigating themselves into position to photograph the storm and—the ultimate goal—its tornado. Many enjoy the thrill of getting close enough to almost feel the tornado winds, yet escaping when the danger is deemed too great. *Tornado Hunter* weaves the fascinating tales of several veteran storm chasers.

Besides capturing tornadoes on video, many storm chasers provide a valuable service as storm spotters, calling the National Weather Service with "ground truth" that there is actually a tornado in progress in association with the rotation that has been spotted by Doppler radar.

A few chasers actually make scientific measurements—either using mobile Doppler radars or with other instrumentation. Their efforts really help the meteorology community learn more about tornadoes and severe thunderstorms.

One of the most remarkable accomplishments in meteorology is to be able to make measurements of a tornado. They are tiny in comparison with most weather systems, difficult to pinpoint, and often fast-moving, and they tend to destroy any conventional measuring instruments in their path. But these hurdles don't stop one veteran chaser—Tim Samaras! He developed special equipment to collect observations in tornadoes, manages to place these instruments in the paths of tornadoes, and then retreats to a safe distance to collect video of the tornadoes in progress. *Tornado Hunter* tells us how he does it and shares tales of adventure along the way.

Tim Samaras is one of the storm chasers I respect the most. I have the deepest respect for his vision and scientific and engineering skills in development of the "turtle" instrument and video instrumentation and for his courage in placing these instruments in the path of rapidly approaching tornadoes. This gives the meteorological community the opportunity to learn things about the near-surface character of tornadoes that are not gathered in any other way, and greatly complements information gathered from mobile Doppler radars and by other storm chasers.

Tornado Hunter intertwines fascinating tales of storm chasing—from some of the most memorable and destructive tornadoes in recent years—with stories of the history of the studies of thunderstorms, lightning, and tornadoes. It describes what weather conditions chasers look for—those that allow thunderstorms to become severe and tornadic. It should be captivating reading for anyone interested in or concerned about tornadoes—especially those who prefer to get the thrill of the chase vicariously!

—*Greg Forbes*

INTRODUCTION

AND GOD SAID TO ELIJAH: "GO FORTH AND STAND UPON THE MOUNT BEFORE the Lord . . . and a great and strong wind rent the mountains . . . but the Lord was not in the wind." These words from Kings 19:11 were spoken at a service held for the small town of Barneveld, Wisconsin, after a devastating F5 tornado struck the community a year earlier, during the early morning hours of June 8, 1984. The spoken words were captured by a documentary camera and became the opening lines for the "Tornado!" episode in the PBS television series NOVA, released in 1985.

The documentary showed meteorologist Howie Bluestein explaining in very simple terms how tornadoes form, and a rather young Louis Wicker (now a research scientist at the National Severe Storms Laboratory) leading a team of brave students trying to deploy a 400-pound data-measuring device called TOTO in the paths of tornadoes. These scientists turned my flame of passion for the weather into a raging bonfire of obsession. Their work showed me that there were indeed scientists actively pursuing research on tornadoes and thunderstorms, and as soon as I could, I became a storm chaser.

Back in 1985, there was no Internet, no cellular telephone, and no 6,000 channels of television. Chasing had a pioneering spirit to it. Unfortunately, though, we often went home empty-handed and frustrated because of the lack of data we could access in the field. Nowadays, one can get a cellular telephone with a modem hookup, then connect to the Internet directly in the field and download the latest radar, surface observations, and satellite and computer model forecasts. With these modern-day tools and good forecasting, our ability to find atmospheric vortices has improved dramatically.

Tornado Hunter takes a peek into a bit of my life, into subjects that I'm extremely passionate about. As a boy, I always wanted to know how things worked. Natural curiosity drove me at a very young age to disassemble television sets to see where the picture came from. Radio also seemed magical when I was very young, and I went on to build several radios made from old television parts. I got my ham radio license when I was 12 years old, and went on to build transmitters that I used to communicate with people around the world. I've always kept my eye on the weather and have enjoyed watching thunderstorms from across the Rocky Mountains. One time, I ran a wire out my bedroom window and attached it to a nearby power pole in hopes of harnessing an actual lightning strike. Inside, I attached it to several lightbulbs to see how many I could light up. Pretty dangerous stuff for a nine-year-old to be doing in his bedroom.

I've never been satisfied with accepting things the way they are, as there is always room for improvement. I've never let anyone tell me that "it can't be done," or that "it's impossible to measure." When people do, it only inspires me to strive harder. After I'd spent nearly 15 years studying tornadoes and chasing them, people laughed at me when I told them I would build a device to actually measure the atmospheric conditions of a tornado's core, improving upon what TOTO had started. I simply smiled and thanked them for their comments. I went on in 1999 to build my first tornado probe, the Hardened In situ Tornado Pressure Recorder (HITPR), which in 2003 measured the lowest barometric pressure drop ever recorded inside a tornado. Little did the naysayers know at the time that I would eventually deploy more scientific instruments in the paths of tornadoes than anyone else on the planet.

INTRODUCTION

In 2007 and 2008, writer Stefan Bechtel endured two storm-chase seasons with us, eating horrible gas station food; going on incredible long, late-night drives; and hanging out with the strangest bunch of people I'm sure he's ever been associated with. There were several times we were quite close to tornadoes, trying to collect measurements, and Stefan was watching us—and the tornado—trying to understand the passion we all share for storms. He certainly captured the thrill that drives us out there every year. His incredible journey with us has been accurately and colorfully written within the pages of this book, and it will take you on a ride that will leave you exhilarated and with a sense of passion and hope.

What is in store for the future? In the 21st century, the holy grail of tornado research is tornado genesis and tornado dynamics. Tornado genesis is basically the birth of a tornado—why do tornadoes form in just a few thunderstorms? Tornado dynamics is the study of the mechanical processes of tornadoes—how powerful the winds are, and other factors. Tornadoes are extremely rare and very fleeting in nature; thus it is nearly impossible to collect data on them.

W. Steve Lewellen, in the paper he presented at the Conference on Severe Local Storms in 2002, suggested, based on his calculations, that winds within a tornado could approach Mach 1 (transonic velocities) for very brief periods of time near the ground. Attempts to measure these high-velocity winds filled with debris using current instrumentation most likely will fail; something new has to be designed. Our attempt will use high-speed photography and other technologies to capture the wind.

The future will bring these and other new tornado probe designs—ones that will accurately measure the three-dimensional wind fields of tornado cores to help us understand how powerful tornado winds really are. Will I have the chance to measure these theoretical wind velocities in tornadoes? I certainly hope so—and please don't tell me I can't do it. I'll simply smile and thank you for your comments, if you do.

—*Tim Samaras*

1: A HOLE IN THE SKY

BY THE TIME REX GEYER GOT HOME FROM WORK THAT DAY, JUNE 24, 2003, SHORTLY after seven o'clock in the evening, the sky had turned a peculiar greenish gray and a few enormous raindrops had begun to splat against the windowpanes. Changing out of his work clothes in an upstairs bedroom at his home in the tiny prairie town of Manchester, South Dakota, he flipped on the radio to listen for weather news.

Thirty-six years old, Geyer worked in materials management for an architectural firm in nearby De Smet, a town made famous by author Laura Ingalls Wilder in her book *Little Town on the Prairie.* He finished changing and went downstairs into the kitchen, where his pregnant wife, Lynette, was heating up some TV dinners for supper. Lynette, too, was listening uneasily to the radio. The local station was tracking an immense and frightening storm system moving across east-central South Dakota.

What both of them were listening for, and hoping not to hear, was news that the roiling atmosphere had given birth to a tornado—the most violent, most freakish windstorm on Earth. In the core of an F5 tornado (see Chapter 10), wind speeds have been documented at 318 miles an hour, the fastest velocities ever recorded on the planet for a natural event. The power of these winds is nearly supernatural. They can reduce a 200-year-old homestead, with all its human memories, to nothing but a windswept foundation in a matter of seconds, hungrily devouring the linoleum from the roofless kitchen floor, as well as the asphalt from the driveway. There was good reason to be concerned. The storm system was bearing steadily to the northeast, the direction everybody knew almost all tornadoes moved. It was the end of June, the time of year when most tornadoes formed. (The springtime months of April, May, and June are prime tornado season.) And it was the early evening, prime tornado time. (Most twisters form in the late afternoon or early evening, after the prairie earth has baked all day in the western sun, stoking thermal updrafts that mount into the magnificent, roiling thunderheads that—if all the other meteorological stars are in alignment—can give birth to a tornado.)

GLOSSARY

Storm system A weather pattern that sets the stage for possible storms

Atmosphere The air around Earth

Stratosphere The part of the atmosphere ranging from approximately 8 to 12 miles

Out here in Dakota territory, where the economy depended on farming, and violent weather was commonplace, people were obsessed with the weather. When a severe storm system like this one boiled up so high it bumped into the edge of the stratosphere and then started bearing relentlessly north-northeast, all other programming was preempted, and the news became all weather, all the time. Announcers would describe the storm's progress county by county, with all the tornado watches, warnings, or touchdowns (if there were any) as they occurred. Farmers, storm chasers, or just ordinary people who found themselves alarmed by what they were seeing would call in to the station on their cell phones, creating a sort of seat-of-the-pants radio CNN, a network of freelance observers who gave listeners an unparalleled account of the storm's progress through the area.

Lynette had a special reason to be worried: Her sister worked in a convenience store in the little town of Woonsocket, 25 or so miles to the southwest of Manchester, which meant that the storm had probably just passed through there.

Anxiously, she called the number at the store, but the phone was out. Which worried her even more.

Lynette was due in three weeks—she was having twins, a pair of girls she and Rex had already named Heather and Hayley—but she'd been having a miserable, problematic pregnancy. On her doctor's advice, this was to have been her first day of full-time bed rest until delivery.

But so far there had been no rest. Now she and Rex sat at the kitchen table listening to the radio announcer as the storm got steadily closer. The thing was bearing down on them, black as soot, towering into the stratosphere, like one of those sci-fi monsters in a late-night movie. Then it started to rain in earnest.

"Why don't you go upstairs and see if the windows are closed?" Lynette asked. Rex went upstairs, and glancing out the bedroom window, he beheld a scene of primal majesty: a black, anvil-headed storm cell swarming across the sky.

Just then the telephone rang. It was Rex's nephew, Wade, who lived nearby.

"Man, you guys gotta look to the south! I think I see a tornado!" Wade shouted over the phone.

Out the bedroom window, Rex could see that the low-hanging storm clouds appeared to be churning or revolving, but he did not see a funnel. There was no lightning. He guessed the storm was perhaps two miles away. It did not appear to be moving at all. Concerned about his mother, who also lived nearby, he gave her a call, but the line was busy. When he glanced back out the window at the storm clouds, suddenly he saw it: an enormous funnel, perhaps half a mile wide, snaking up into the low-slung cloud deck or belly of the storm. Apparently, the storm had passed into a cornfield, and the previously invisible vortex began sucking up black dirt, abruptly making itself terrifyingly visible.

Mark Strickler, a former wheat farmer who lived across Highway 14 about a mile south of Manchester, was also listening anxiously to the radio that evening, glancing out the window at the glowering sky as he fixed his supper. He could see a cottonwood tree heaving over in the wind, its leaves turning up their silvery undersides. The storm clouds were black now, racing across the sky. Then he saw it—a small, almost mischievous funnel of wind, not much bigger than a dust devil, spinning down into a field about a half mile from his house. It danced a short distance across the grass, then seemed to spin apart, dissipating back up into the low-hanging ceiling of dark storm clouds. He stepped out onto the porch with his plate in his hand and watched as another little weak, wavering tornado dropped down into a field, skipped merrily along the grass, then disappeared, as insubstantial as mist. Then he saw another little tornado, then another.

Mark had lived in South Dakota his whole life, but he had never seen anything like this. Sure, these tornadoes were all frail and fleeting, shadow dancers, more like the eddies in a gentle stream than the roaring cataracts of white water. But actually witnessing a tornado once was relatively unusual—much less four or five of them in ten minutes.

That's when he saw it. The menacing storm clouds racing by overhead coalesced into something entirely more frightening than a mere summer thundershower, no matter how black and fearsome. Now he could see them revolving into a great, dark vortex that seemed about to swallow the world. The funnel dropped all the way to the ground and began racing toward him across the wheat fields, abruptly lifting and then passing directly overhead. Mark found himself standing there on the back porch, with his supper plate in his hand, staring up inside the weird, alien core of a real live tornado. Later he would say that it looked like "a hole in the sky," a great black whirlpool with lighter, roiling storm clouds up inside. It was such a terrifying moment that retelling the story four years later, this enormous, laconic man, a midwestern farmer, had tears in his eyes.

"I hope I never see anything like that again," he said.

Although the dark vortex passed over his house without incident, now it was bearing down on the little town of Manchester. The enormous funnel was tearing up the fields, lifting up the black Dakota dirt into the air in a great, eerie cloud of chaos and destruction.

It was headed straight for the home of Toby Towberman, the man everybody considered the unofficial "mayor" of Manchester. Toby Towberman was a big man, tipping the scales at more than 300 pounds, with a laugh to match: People said his rollicking roar could be heard for three blocks in any direction. He made for a slightly comical figure, this immense, genial, top-heavy man, tottering around town, always followed by a little white lapdog named Cody. Toby's kindnesses were as outsize as his body. One time when a caravan of cars got stranded in snow along the highway, he and his wife, Kris, took all ten people into their little house, fed them dinner, and let them stay overnight. That was just the sort of thing Toby was always doing.

At one time, being mayor of Manchester might have conferred some actual pay, or at least worldly status, because at the turn of the century Manchester had been a real town, with 50 or 60 full-time residents. A photo postcard dated November 22, 1910, which cheerily sent "Greetings From Manchester, South Dakota," depicted a cinematic western town complete with a grain elevator, a church, and a dirt main street lined with sleepy saddle horses and false-fronted buildings. Once, in 1961, a local TV station had arranged a gala Dakota Territory Centennial Celebration in Manchester, featuring North Dakotan Lawrence Welk and a young actor named Clint Eastwood, newly famous in his role as Rowdy Yates on the TV show *Rawhide.* The event drew an incredible 150,000 people, mainly because the TV station, KELO, hid little sacks of gold in a few of the thousands of three-by-three-foot "claims" that people could buy as part of the event.

The Dakota Territory Centennial was the second to last time the town was famous, because after that, like so many other tiny Great Plains towns, Manchester gradually blew away like dust in the prairie wind. Lots of people had already gone bust and moved away in the Dust Bowl days of the 1930s. Then, the steam locomotives no longer had to stop for water in every little town. Life on the prairie, always tough, became tougher, and people slipped away to bigger towns and cities. Manchester's population had dwindled to only six people by that evening of June 24, 2003, when a black whirlpool of wind spun down out of the sky and the little town briefly became famous one final time when it was wiped off the face of the Earth forever.

Mark Strickler could see that the tornado was so wide it seemed about to engulf the entire town. He prayed that Toby and Kris weren't home, or at least that they had heard that the tornado was coming and found shelter. The funnel crossed the highway, and then Mark saw Toby's house simply lifting into the air "just like that scene in *Wizard of Oz.*" He could see the house, spinning, being taken aloft into the clouds. He could see a few shingles or boards come loose. And then the whole house just exploded, spinning out into a ragged shroud of debris, hundreds of feet up in the air.

Toby and Kris were, in fact, home that day. They'd heard about the oncoming tornado on the radio, and as the sky turned black and the wind began to shriek in the eaves, they headed for the basement. In tornado country, such decisions are generally made in extreme haste. Kris ran down into the basement, but Toby, overweight and with bad knees, moved much more slowly. He heard his beloved little dog, Cody, locked out in the garage, scratching frantically at the door. Toby let the dog into the house, then bolted for the basement door. Kris, already down at the foot of the basement stairs, reached

A tornado snakes toward Tim Samaras and his tornado-chasing van.

up to take her husband's hand. Toby had made it down to the third stair into the basement and was reaching out for her when suddenly he, the dog, the kitchen, the roof, and the entire house were lifted into the sky.

Kris's uplifted hand was empty.

Rex and Lynette heard somebody roaring up into the driveway. It was Rex's brother Dan, with his wife and two children in the car.

"Man, we gotta get in the basement!" Dan yelled, charging into the house. *"Just look at that thing!"*

Rex suddenly realized why the tornado appeared to be motionless, not moving to the left or to the right. It was because the funnel was headed directly for his house. And now he could see that the oncoming storm was filled with flying debris—not just shingles and leaves but dimensional lumber, two-by-fours, and other structural elements that meant it was grinding buildings to pieces as it approached.

Six of them were standing there in the kitchen as the tornado bore down on them—Rex, Lynette, Dan, Dan's wife, and the two children. Dan had no basement in his house, but Rex's house was not a great deal better—it was just an old prairie farmhouse with a rough, rock basement, not too deep. Plus it contained two fuel tanks, including one filled with 500 gallons of diesel. If that twister hit the house, they could all be dragged out and killed. There was also the great danger that the fuel tanks might explode.

GLOSSARY

Tornado Alley The nickname given to the south-central United States, where a high frequency of tornadoes occurs every year

"I don't think it's safe in the basement!" Rex insisted. "I think we need to get outta here!"

Dan had parked the car, engine running, right outside the front door, so the six of them piled out of the house and into the car.

"Should I turn off the lights and the TV?" Lynette asked Rex weakly. He was always bugging her about that.

There was almost no visibility outside. The rain was coming down so hard that even with his wipers on high, Dan could barely see where he was going. Then the pounding rain turned to hail, hammering on the car like gunfire.

"What should we do? Where should we go?"

Rex and Dan were yelling back and forth over the roar of the rainstorm, but neither one knew exactly what to do. It was often said that you should never try to outrun a tornado. For one thing, tornadoes can travel amazingly fast—often more than 30 miles an hour, and in one famous case, more than 70 miles an hour. They can simply outrun you. And being trapped in a car while a twister overtook you was a possibility almost too frightening to be imagined.

Dan took off to the north, along a hard-packed gravel township road. North seemed to be away from the storm. But the tornado seemed to be bearing down on them, though it was hard to see anything at all. Around them was just blackness and the pelting din of hail.

"Turn left! Turn left!" Rex shouted. Dan wrenched the car left, bearing west on another county road. They came by Harold and Loretta Yost's house, barely visible in the storm, and Dan lay on the horn, hoping to warn them if they were home. (Luckily, they were not.) About a mile to the west, they seemed to emerge out of the darkness and hail of the storm. Glancing back, they could see what looked like a big, angry summer thunderstorm, throwing down slate-colored veils of rain. They couldn't see any tornado at all. They turned south on another township road, then back onto Highway 14, bearing east toward Manchester. Abruptly, there it was again: an enormous funnel cloud. It seemed to be churning straight up Main Street, almost due north, right through Manchester.

Rex could see that Toby Towberman's house was completely gone. He could also see that the road around Manchester was crowded with vehicles, several of them with strange-looking contraptions made of PVC pipe mounted on the roofs, somewhere between modern art and plumbing.

Two vehicles flashed past Dan Geyer's car in the rain—a white van bristling with antennas and electronics gear, followed by a red SUV. Piloting the white van was Tim Samaras, a Denver-based inventor and engineer who was probably the most ingenious and persistent tornado researcher in the world. During the summer months of May and June, Tim and his team ranged across the high plains of Tornado Alley, attempting to deploy a small disk-shaped instrument probe directly in front of a tornado. The goal was to take measurements of barometric pressure, temperature, wind speed, and humidity from the forbidding interior of the tornadic core, to understand how tornadoes form, how they gain power, how they move, and how they disintegrate. But the ultimate goal of this scientific adventure was to help protect human life—shielding somebody's mother, brother, sister, or daughter from the naked savagery of Earth's most violent windstorms.

Today Tim was traveling with his storm-chasing partner and brother-in-law, Pat Porter. Behind Samaras and Porter, in the red van, was veteran storm chaser Gene Rhoden; his Scottish wife, Karen; and German photographer Carsten Peter, on assignment for *National Geographic* magazine. Carsten's mission: to capture extreme close-up images of tornadoes, a task almost as difficult and dangerous as Tim's attempts to deploy a probe in front of them.

A rangy, long-legged adventurer with a perpetual five-day growth of beard, Carsten looked a bit like actor Viggo Mortensen as the gaunt warrior Aragorn in *Lord of the Rings.* When Carsten came over from Germany on assignment for the magazine, he and Tim had hit it off immediately; there was no need to explain the avid pursuit of something that was darkly mysterious and phenomenally dangerous, and whose understanding would be of enormous benefit to science and humanity. Carsten understood.

For Carsten, the fear and chaos of a tornado touchdown was pure ecstasy. He lived for these moments. In fact, his whole life was arranged to be present, camera in hand, when such moments occurred. He was, he said, "addicted" to the grand spectacles of nature, especially volcanoes, having photographed their steaming calderas and gaseous, otherworldly ash plains all over the world. He once mounted a famous 55-person expedition into a volcanic crater on the island of Vanuatu, in French Polynesia. He'd penetrated bat caves in Mexico, Franz Josef Land in the Soviet Arctic, and the Kamchatka Peninsula, rambled from the Sahara to the high-elevation rain forests of Peru and Bolivia. But for someone bewitched by nature's drama, tornadoes offered something with an almost incomparable allure.

A HOLE IN THE SKY

In many ways, Tim Samaras and Carsten Peter were the luckiest of men. They were among the fortunate few whose profession was their passion, whose avocation was their vocation. Living in that "sweet spot" where work, play, and joy were all the same thing, they exuded a kind of perpetual youthfulness, as if their energy were inexhaustible. They had worked out agreements with life such that they had the freedom to spend at least two months of the year roving thousands of miles across the high plains of the western United States like wild Indians, seeking not only tornadoes but also the ecstatic high of those luckiest of days—those days when nature seems to disclose her most closely held secrets, however briefly, and time stands still.

This day—June 24, 2003—was about to turn into one of the luckiest of those lucky days.

2: OUR PROBE'S STILL THERE!

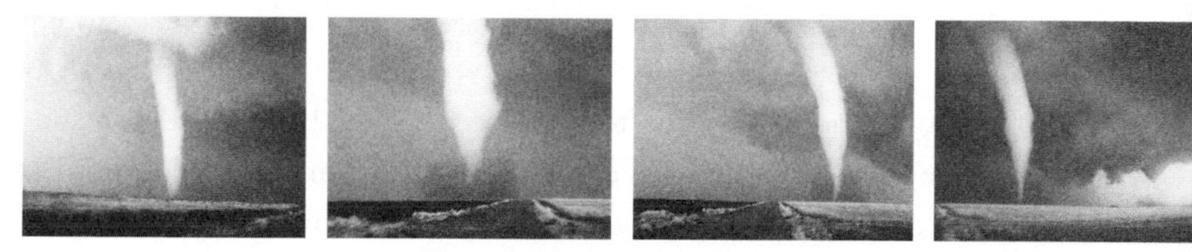

IT HAD ALREADY BEEN AN UNFORGETTABLE DAY FOR TIM SAMARAS AND CARSTEN PETER. The two of them had worked together deploying probes and taking photographs for three storm-chasing seasons, logging more than 30,000 miles a season, chasing storms to as far north as southern Manitoba, east to Iowa, west to Colorado, and all across the Dakotas, Kansas, Nebraska, Oklahoma, and Texas. But so far—at least in terms of coming face to face with a twister—the enterprise had been a bit of a bust. They'd seen plenty of movingly beautiful prairie storms, lots of supernatural lightning and thunder, and the inside of more than their fair share of Super 8 motels. But they'd spotted only a few weak, brief, or distant tornadoes—nothing near the terrifyingly close-range, highly cinematic, and photographable one Carsten was longing for.

The primary goal of Tim Samaras's work out here was not merely to photograph a tornado at close range. It was to get *directly inside* the tornado's core and take scientific measurements there, especially of the lowest 30 feet—one of the most mysterious and poorly understood regions of the tornado.

"That lowest ten meters [30 feet] is incredibly chaotic and, of course, dangerous, and it is just very difficult to get any information about it," Tim says. "Also, mobile Doppler radar can get good information about higher elevations of the tornado, but not down close to the ground. As a consequence, we still know very little about it.

"Why does this lower region matter? Because that's where we all live."

The core of an EF4 or EF5 tornado (see Chapter 10) is as freakish and hostile a place as the atmosphere of Mars. Though most tornadoes last 15 minutes or less, the phenomenal wind speeds and pressures at the center of the cyclone make it a place where the laws of physics of ordinary Earth life seem to be repealed. Cars, rooftops, and animals become weightless and take flight; tractor trailers tip over like Matchbox toys; ancient oaks are plucked whole out of the ground and levitated. In the tornado's aftermath, oddities abound, such as phonograph records rammed sideways into telephone poles; full place settings left undisturbed on a dinner table in a house without walls; or, in one well-documented case, near Kickapoo, Kansas, a man set down alive after being lifted up and carried through the air for more than a mile. (He later died.)

To actually penetrate the core of a tornado by deploying a probe on the ground and have it emerge with real information is a task so difficult it's akin to trying to take the temperature of a *T. rex*. Even getting a basic measure of tornado wind speeds is an enormous challenge because tornadoes simply obliterate anemometers put in their path. It was not until 1958 that anybody actually got indirect measurements of wind speeds inside the core using Doppler radar. It was not until 1999 that the first wind measurement in excess of 300 miles an hour was made by tornado researcher Josh Wurman, with a truck-mounted Doppler radar, but these measurements were not at the ground.

The idea of developing some kind of armored probe, loaded with instruments, that could actually *get inside* the whirlwind and take measurements of the lowest 30 feet was repeatedly attempted, but never successfully. (That was the premise of the 1996 hit movie *Twister,* a fanciful interpretation of actual experimental probes under development at Oklahoma University and elsewhere.)

Partially funded by a series of grants from the National Geographic Society, Tim Samaras kept refining his design of a circular, instrument-laden, armor-plated probe that he called a "turtle" or sometimes simply a "cool orange hat." The current design of his probes, which weigh about 45 pounds apiece and are roughly the size of an extra-large manhole cover, slope to a point at a 30-degree angle, so that they look something like a Chinese peasant's hat (albeit one made of steel). Though the shape might seem whimsical, it is aerodynamically ingenious, because when Tim tested various configurations in a wind tunnel, he found that in the teeth of screaming 225-an-hour winds, this shape was pushed *down* by the wind rather than lifted *up.*

"We used the wind to our advantage," Tim says simply.

Unlike, say, a Mars rover—another kind of robot probe used to explore a hostile and alien environment—Tim's tornado probes don't actually move. In fact, the whole idea is *not* moving: They just lie there, motionlessly hugging the ground like a lamprey stuck to a shark's back, taking measurements of the tornado via instruments tucked safely inside the hardened shell while all hell breaks loose outside it.

They are painted a brilliant, cheerful orange, so as to be easier to find in the rubble-strewn aftermath of a tornado strike. Inside the hardened shell are instruments to measure the tornado's wind speed and direction, barometric pressure, humidity, and temperature.

The probe's more humorless, scientific name is Hardened In situ Tornado Pressure Recorder, or HITPR. The HITPR probe is, in effect, an armor-plated mobile weather station and surveillance robot.

The HITPR probes are also living proof of the special skills Tim Samaras brings to the task of understanding tornadoes. Though he himself does not have a degree in meteorology, he has talents that most meteorologists lack, such as a lifetime of experience as an ingenious tinkerer and a strong background in engineering. At Applied Research Associates, Inc., an Albuquerque-based engineering firm, he has every ten-year-old boy's dream job: blowing up stuff and then studying what happens. He knows about designing things to withstand enormous impacts and explosions. He was able to test his "cool orange hats" in a wind tunnel and determine the precise shape that would create an eddy in the tornadic winds to ensure that the instruments could survive.

GLOSSARY

Doppler radar A radar that measures the instantaneous motion toward or away from the radar beam

Cumulonimbus Dense, detached clouds with sharp outlines that develop vertically into huge towers due to unstable air; they are often called thunderheads.

Wedge tornado The slang term for a tornado whose funnel is at least as wide at the ground as it is tall (from the ground to the cloud base).

Tim's mission, as always, was to place an HITPR probe directly in front of a tornado, or at least somewhere close to the path of one. The task appeared to be so difficult that plenty of severe-weather experts felt that the whole idea was a waste of time; the data you might gain came at too high a price. But Tim Samaras did not believe the experts, and he just kept on trying.

Now it was the end of Carsten's third season of storm chasing, vying with the elements for the prize—an up-close image of a major tornado—and *National Geographic* was nearing the end of its patience. Carsten had called to beg his editor for just one more week in the field, and the editor agreed to free up the money for seven more Super 8s and some road food and fuel. But no tornado. Then, yesterday, Carsten had called his editor one more time, begging for another day—*just one more day!* Everything seemed to be shaping up beautifully, he promised. The weather looked extremely promising—that is, really, really bad. All the complex prerequisites for tornado formation were falling into place.

They were in the right place (the Dakotas) in the right time (June). There was warm, moist air out of the Gulf of Mexico pouring north across the great plains, colliding with cool, dry air pouring in off the Rockies (one of the key reasons why the conditions are ripe for tornadoes in the Great Plains of the United States more often than any other place on Earth). At the same time, there was high temperature and humidity, so that the dew point (the temperature at which water will condense) was at almost 70 degrees, meaning that the air was absolutely loaded with moisture. This meant that there would be immense cumulonimbus clouds piling up to the edge of the stratosphere. Finally, there was also high-elevation atmospheric instability and directional wind shear, spiraling around as it gained elevation, which could cause a boiling thunderstorm to begin to rotate. It was a situation, Tim would later say, in which "everything that went up, rotated."

Prime tornado weather.

So *Geographic* editor Todd James granted Carsten one last day. And finally, at last, it had happened. Just an hour before, he and Tim and the rest of the team had witnessed and photographed a monstrous tornado, eerily enveloped with shrouds of dust and rain, near the town of Woonsocket, South Dakota. (One of Carsten's pictures would become the cover photograph of the April 2004 issue of *National Geographic.*)

Now they were tracking the vast storm system that had spawned the Woonsocket tornado to the north and east, bearing down on the town of Manchester. Not far from where they were, about a mile north of town, a farmer named Jim Bowes stood outside his house watching as an immense dark wall cloud swept toward him. Suddenly, out of the cloud bank, the funnel itself emerged, snaking up into the sky. For 15 or 20 seconds, the funnel was "a bright, neon lime-green," he would recall later. "It just glowed." Whether this was because the twister had taken aloft an enormous load of green vegetation (afterward, the sides of Bowes's silos were splattered with green even though his barns survived intact) or because of some other optical effect, its appearance was utterly eerie, unlike anything Bowes had ever seen or even heard of.

Moments later, Tim Samaras, Carsten Peter, and Pat Porter witnessed the great sinuous horn of the tornado emerging out of the muddy darkness and rain veils of the storm. The tornado was churning across a field to their right, perhaps a quarter mile from the road. It appeared to be moving steadily toward them.

"Wedge tornado on the ground!" Tim yelled.

"Oh my God!"

Carsten Peter, following in the red SUV, hung out the open window, firing away like mad with his Nikon F4. He could see the thing plucking up telephone poles and flinging them into the air like toothpicks. It all made him feel supremely alive and half crazy with glee. These were the moments he lived for. What a day this was!

"Unbelievable! Unbelievable!"

The thing before them was eerily, transcendentally beautiful, moving across the green-golden fields with a sort of majestic stateliness. It had taken the shape of an immense mushroom, with a huge gray pillar a half mile wide mounting up to the low-hanging cloud deck. Around the spinning core, turning in ever slower revolutions the farther they were from the center, were gray shrouds of dust, debris, and water vapor.

Its beauty, its mystery, and its danger were all spinning out from the same core.

Tornadoes, like all atmospheric phenomena, are continuously shape shifting, but at this moment it was wider than it was tall (hence the term "wedge" tornado). Wedge tornadoes can be particularly devastating, since so much of their vortex is making direct contact with the ground.

While Pat kept the video camera running, capturing everything as it unfolded, Tim piloted the van down a rain-slick township road, trying to get closer to the twister so he could deploy a probe directly into its path.

"Man, this is too close for me—I am not going in there!" Tim said uneasily.

He stopped the van, and he and Pat sat there, watching the thing. Though they were less than a quarter mile from something that could suck them both up into the sky, and the SUV along with them, their voices, captured on the videotape, were remarkably calm.

"Are we too close?" Pat asked.

He turned north down another township road, following a street map program he was getting using GPS technology, which showed him the entire grid of roads, even hard-packed gravel roads, in the area. It was useless to locate a tornado if you couldn't get close enough to deploy a probe into it, and the only way to get close to it was by following the available roads. Now they stopped again, watching as the tornado bore down on a small copse of trees with a small house and outbuilding nearby. The storm was so immense it seemed to miniaturize the cattle, buildings, and vehicles far below. On the video, the tiny outbuilding can be seen tipping over and its roof coming off, like a woman's hat flying off in the wind. Then the outbuilding collapses and debris flies straight up into the air. Tim and Pat watched as the roof of the house collapsed; then pieces flew directly up into the air and the whole scene seemed to disappear into a maelstrom of dust and debris. Then Tim and Pat continued north to outrace the tornado.

"It's a dead hit!" Pat said, in awe, as the buildings swirled up into the sky in a gyre of leaves. Now, abruptly, the storm enveloped them and the tornado seemed to disappear.

"We're losing visibility," Tim said as he continued driving north, recognizing the danger of the situation. There was still a violent vortex in extremely close proximity, but they could not see it. Now, in the muddy chaos of the storm, it was difficult to tell how far they were from life-threatening peril.

Carsten Peter also recognized the precariousness of the situation and was rapidly trying to assess the risk. He lived a life most people would consider insanely dangerous, having come in close contact with poisonous volcanic gas and lava as routinely as most people take a vacation at the shore. He'd contracted malaria seven times, and nearly died from it more than once. But he was calculating about his approach to any situation. It wasn't that he was recklessly in love with or courting danger; being around it so much, he had learned to manage risk as a professional hazard. (He had also chosen not to marry or have children, at least partly so that the risks he chose were his own alone.)

One might say that both Tim Samaras and Carsten Peter shared a curious kind of kinship, in that they had both learned the *mastery* of danger. They knew how to quickly size up a situation and to calculate the risk, because without risk they could not do what they wanted to do. They knew how not to become ego involved or prideful, and how to withdraw unapologetically when the risk was too great.

Getting this close to a tornado, especially with bad visibility, was the sort of situation that could turn deadly in a matter of seconds, and both Carsten and Tim knew it. It wasn't like you got "pre-warnings" about what was going to happen next; you had to anticipate things, which involved second-guessing the movements of one of the most unpredictable and violent phenomena in nature. Unless you were very experienced, knew the territory, and had real-time weather and road information, doing this was crazy. It was irresponsible. You could very suddenly find yourself trapped, by a downed power line or a tree across the road. You could run off the road and get stuck in the mud. It was so easy to lose cell phone contact with the other chase vehicles, because the cell towers could blow down in the storm. Pelting rain, golf ball–size hail, and debris could cut visibility to near zero. Suddenly, you could find yourself trapped and alone, in a situation that was out of control. It was like being trapped in a cage with an angry bear—in fact, the most dangerous part of a tornado was often referred to as the bear's cage.

And it could all happen in the blink of an eye.

OUR PROBE'S STILL THERE!

Even so, Tim kept trying to pilot the van closer and closer to the storm.

"Man, I'm not sure we have time," Pat told Tim uneasily. "This is too dangerous."

As the tornado came boiling across a field toward the chase team, completely filling the video frame in front of the vehicle, Tim stopped the van and climbed into the back, to retrieve and deploy one of six probes he had in the vehicle. *"We don't have time!"* Pat yelled. *"We don't have time! Seriously!"*

But Tim threw out a probe beside the road anyhow, perhaps 200 yards from the tornado, and then they took off, racing down the road away from the storm. They seemed to emerge from the confusion of the rain clouds, and then stopped again at a country crossroads—two flat, straight South Dakota roads intersecting squarely, with a stop sign marking the intersection. Around them the sky was black; nearby, golden-green fields of grass were undulating wildly in the wind. The tornado, now changing from a wedge to a gray, steadily narrowing funnel, with great clouds of dust and debris revolving around its base, still appeared to be about a quarter mile away.

Behind them, the tornado was clearly visible, now shape shifting into a great smoke-colored python, snaking up into the sky with clouds of debris, like a dancer's skirt, revolving around its base.

"Just went through that house!" Pat said, watching another building being sucked up into the sky.

"OK," Tim said, "I want to stop and look at it!"

He stopped the van and got out. Behind them, Gene Rhoden's red SUV also stopped, and Carsten climbed out.

"Oh my *God!*" Carsten yelled, jumping around in the road. He was ecstatic. "Oh my *God!*"

"Listen to it!" Pat said.

"It's coming right at us, too! We have to put another probe out!" Tim yelled.

Tim ran back to the van, manhandled another probe out of the rack, and raced down to the edge of the road to deploy it. It was the third deployment of the day. Behind him the sky was getting steadily blacker. The dust cloud at the tornado's base seemed to grow blacker, wilder, more malevolent, piling up into the sky as if the world were ending in black smoke and fire.

"That's gonna miss us, Tim!" Pat yelled. The tornado seemed to be bearing steadily to the east, moving away from the road and the two deployed probes, its gray funnel snaking into the sky. People often describe the roaring sound of an approaching tornado as akin to a freight train or a jet engine, but the roaring they could hear now was entirely inhuman, and of almost unimaginable intensity. The air began to fill with dirt, debris, and swirling vapor. Then it seemed to pause, slow down, and veer back toward the road. Moments later the tornado was bearing down on them. It was perhaps 200 yards away, and closing in.

"Let's go, Tim! *Let's go! Whoa!*"

They jumped back in the van and took off again, then stopped, watching. They could see it moving toward what looked like a copse of trees but was actually the little town of Manchester, in its last few seconds of existence.

Pat: "I want to see it go through those trees, Tim."

Tim: "All right."

They watched it bear down on the trees, perhaps a quarter mile down the road.

Pat: "It's coming back on the road! Oh, it might've hit your probe!"

Then the tornado veered off the road again to the west.

Pat: "Probe! Let's deploy another probe!"

Tim: "Have we got time?"

Pat: "Yeah, it's not at the trees yet!"

They stopped to deploy another probe on the west side of the road as the tornado approached them.

"The money shot!" he yelled, joyously. "Is this outta control, or what?"

They jumped back into the van, racing a short distance, then stopping to get out and observe and take pictures. In the van the local radio murmured continuously, an undercurrent of alarm: ". . . tornado touchdowns in Kingsbury County . . . please take cover . . . moving east at 35 miles an hour . . . very, very serious storm. . . ."

Now the tornado took on a new shape. The thick, gray funnel seemed to grow translucent, and a much narrower, whitish, undulating serpent appeared inside the previous funnel, twisting up into the sky.

"Rope stage!" Tim announced. "Look at that laminar tube!"

Now the rope or laminar tube grew increasingly skinny and seemed to separate in places, like fibers of cotton being yanked apart by some unseen hand. Then the funnel rapidly faded away completely, leaving only a muddy, clearing sky and a mischievous swirl of departing dust, harmless and disorganized.

Tim got out and stood beside the mud-spattered van, holding his arms aloft.

"Pieces of insulation falling down," he observed to no one in particular, as all around him the sky rained debris from what once had been home and hearth to the people who lived here.

Tim and his team drove back to the place where he had dropped the first probe. The devastation in Manchester was complete. The crew stopped the vehicles and got out, walking through the freshly ravaged landscape, marveling at it all, taking photographs and video. A small collection of other cars were already there—the sheriff, as a first responder checking to see if emergency medical help was needed,

GLOSSARY

The bear's cage The precipitation that wraps around a mesocyclone and might hide a tornado

Tinman A pyramid-shaped aluminum casing loaded with a video and three 35-mm still cameras designed to take pictures from a tornado's core; it takes its name from one of the characters in *The Wizard of Oz.*

and other members of the small, slightly crazed community of storm chasers. One of them, Sean Casey, stepped out of what looked like a gray submarine on wheels, but was actually a custom-designed Tornado Intercept Vehicle—a Ford F-450 Super Duty truck encased in one-eighth-inch steel, with inch-thick Lexan polycarbonate windows. Sean was attempting to film a large format film about tornadoes from as close as humanly possible (so far, after years of trying, with little success). The TIV was essentially a tornado probe, with a human inside it.

"You OK?" Tim asked Sean.

"Yeah, man. How 'bout you?"

For the first few minutes of the aftermath, Tim and his team became first responders, scouring the wrecked houses for any sign of life before turning their attention to science. Fortunately, there had been only a few outbuildings and a handful of inhabitants—but all the buildings had been raked open and completely obliterated. The foundation of one house stood gaping open like a disemboweled tomb, huge fieldstones scattered recklessly about, as if earth-giants had been afoot. All the trees were either ripped out of the ground by the roots or, if they were still upright, scoured clean of leaves and small branches. Bare limbs reached up to the sky like penitent, praying hands.

"This is F5 damage," Tim said, looking around. "F5 or F4."

Ted Fujita, the brilliant Japanese researcher who was the developer of the Fujita scale of tornado intensity (see Chapter 10), described F5 damage simply as "incredible." F4s produce "devastating" damage. (Later the storm would be classified as an F4 tornado, with winds in excess of 200 miles an hour.)

Carsten, who had come equipped with a steel-encased photographic probe known as the Tinman, had succeeded in setting the probe beside the road a few minutes earlier. Embedded with cameras, the Tinman was an attempt to get photographic images from inside the core, a place that had never been photographed before. But Carsten had been unable to find the probe when he went back. The storm had sucked it up into the sky. The design comparison between Tim's squat, ground-sucking UFO probes and the exposed, upright Tinman was painfully clear.

"Have you seen my photo probe?" he asked the sheriff plaintively. "It's a silver triangular device along the road somewhere—please let me know if you find it."

The Tinman had gone missing. (A day later the Tinman would be found by the team in a field about 300 yards away from the spot where Carsten placed it, smashed and useless. Even so, Carsten's award-winning photographs from this close encounter would become not only *National Geographic*'s cover story in April 2004 but also its first successful tornado story.)

For the former residents of Manchester, photographic firsts or scientific discovery were of less interest than flesh and blood. One by one, the survivors began showing up at the scene of destruction. Mark Strickler arrived in his pickup, his face a mask of worry. He found Toby Towberman's wife, Kris, seated on the basement stairs of what had once been her home. All that remained was a cement foundation, swept almost completely clean. It was as if the house had never been there at all.

"Where's Toby?" Kris asked, her eyes glazed and uncomprehending.

"I don't know," Mark told her.

He could see that the rubble was strewn to the north and the east, so he followed the tornado's contrail of destruction, shouting as he went.

"Toby! *Toby!* Are you there? Are you OK?"

Finally he heard a muffled groan and found Toby Towberman, the mayor of Manchester, sprawled at the base of a knocked-over cottonwood tree about a city block away from the ruined house. Around him were the remains of his kitchen floor. In the branches above him was the crushed form of his beloved little white dog, Cody. The dog was dead.

"Where am I?" Toby moaned. "Where's Kris?"

"She's all right, Toby—she made it. Are you OK?"

"Where am I?" Toby repeated. "Where's Kris?"

Mark Strickler took Toby's bloody head and shoulders into his arms and cradled him like a big bear. A few minutes later, when a woman from the Red Cross found him standing nearby, she grew concerned that he was covered with someone else's blood. In the age of AIDS, human blood could be poison.

"Lady," Mark told her, "this is Manchester. Everybody knows everybody. If Toby had anything, I'd have it already!"

Rex and Lynette Geyer, Rex's brother Dan, and Dan's wife and children drove back to Rex's house. All that was left was the gaping hole of a bare foundation, with two huge fuel tanks exposed (though unexploded.) The house had been completely swept away. If they had sought shelter here, they would almost certainly have been killed.

"There it is!" Pat shouted. *"Guys! It's still there! Our probe's still there!"*

For such a stellar scientific moment, the probe itself looked remarkably innocuous—a cool orange hat lying beside the road, completely slathered with mud and gravel. The Tinman might have gone missing, but the humble turtle had lived through the tornado. Tim and the others circled around it, photographing the probe in situ (in place) before anybody touched it, as if it were about to explode. Then Tim knelt down and tipped it over to check: The little green light, signaling that it had been recording data, was still bravely blinking. The tornado had scoured the gravel road completely clean, but when Tim lifted up the probe, underneath was a little circle of gravel, entirely undisturbed. The turtle had stuck to the road like a lamprey, just as it should. Tim Samaras had succeeded in doing something nobody in the world had ever done before—designing a probe that could survive the most extreme atmospheric stresses on the planet, then deploying it directly in the path of an F4 tornado.

And it had come through alive and virtually unharmed. Now all that remained was one small detail: to see if it had actually worked, recording usable information about the tornadic core.

But Carsten wasn't going to wait for that. His reaction was one of pure joy.

"It's amazing! It's amazing! It's amazing!" he shouted, jumping around in the road.

Tim, as usual, was calm and level, even after what appeared to be an astounding scientific achievement. He had Carsten take a few pictures of him kneeling beside the probe, then tilting it up off the ground to show the untouched circle of gravel underneath. But until he had downloaded actual data, he was not about to break out the bubbly.

Later that evening, in a cheap hotel room, the moment for celebration finally arrived. "When I downloaded the probe's data into my computer, it was astounding to see a barometric pressure drop of a hundred millibars at the tornado's center," he said. "That's the biggest drop ever recorded. That's like stepping into an elevator and hurtling up a thousand feet in ten seconds." It's a drop of such magnitude that some atmospheric scientists had argued it was not possible—at least, not until now.

When he later plotted the barometric pressure readings from the probe onto paper, one could see a relatively flat line, tilted slightly downward, meandering across the page (representing the outer wall of the funnel). Then there was a sudden, precipitous drop, as if the line had fallen off a cliff (representing the probe's entry into the extreme low-pressure field of the tornado's core). The amount of time that elapsed from the beginning of the graph (when Tim switched on the probe and flopped it on the ground) to the "cliff" (the edge of the cyclone's center) represented 82 seconds. Tim Samaras was kneeling in the road 82 seconds before an F4 tornado passed over the exact same spot. Thirty feet away from that spot, a hundred-year-old house was reduced to rubble.

As Tim and the other storm chasers poked through the sad wreckage, checking to see if there was anybody who needed help, miles away, across the fields, yet another tornado was visible, a slate gray funnel descending like a staircase to the ground. But they barely seemed to notice. It had been a once-in-a-lifetime day for Tim Samaras as a tornado hunter. His team had logged hundreds of thousands of miles and years of effort tracking these things in order to extract secrets from the core, often to no avail, but on this single day not only would they capture the most astonishing photos ever made of a tornado, but they also would record the greatest barometric pressure drop—a scientific first. While almost 70 tornadoes would be reported in South Dakota that day—an all-time record—Tim's team would encounter and document 7. But none of the others could compare with the F4 that destroyed Manchester, in the process allowing one fleeting glimpse into its very core.

The town of Manchester, all half dozen buildings of it, was completely leveled; its long, bittersweet, human story had seemingly come to an end. In the fields outside the devastated town, livestock bellowed, disturbed and frightened. Later it would emerge that several dozen cows had been killed, others thrown over fences, their backs and legs broken. The perverse and mischievous "intelligence" for which tornadoes are famous was in evidence everywhere. One farmer would later find a half mile length of barbed-wire fence, ripped out of the ground, its fence posts popped off, the remaining wire rolled up into an enormous, spooky, circular bale in the field. And though the video clearly showed the twister turning counterclockwise (as virtually all tornadoes do in the Northern Hemisphere), a telephone line would later be found snapped in two and tightly wound around a telephone pole *clockwise.*

"One of the most amazing things about the Manchester tornado was that, even though this was a truly ferocious beast, an F4, nobody got killed," Tim says. "And we're always thankful for that."

In the distance, out across the great prairie grasslands, the underbelly of the sky had already begun to brighten after the storm, as if someone were opening the lid of the world. You could hear the sweet, incongruous warbling of a meadowlark, often called a tornado bird because its is the melodious coda one hears after the last act of a storm.

It was almost as if nothing had happened at all.

3: LOOKING FOR TROUBLE

TIM SAMARAS DOES NOT APPEAR TO BE A GUY WHO'S LOOKING FOR TROUBLE. BUT here he is, on this sun-drenched South Dakota summer afternoon, trying very hard to expose himself to grave bodily harm. He wants to go find the whirlwind and then stick his nose inside it. He is on precisely the opposite path from almost everybody else in this world. What other people call bad weather, he calls good weather. Wherever they are fleeing from, that's where he's going. If they are running from absolute disaster—houses sucked into the sky, ancient trees snapping like matchsticks—that's when he knows he's exactly where he wants to be.

At this moment he's piloting the black, specially adapted all-wheel-drive GMC truck that his three children call the geekmobile, out of amused fondness for their dad. Storm chasers typically love gadgets and gizmos dearly, but Tim's geekmobile has them all smoked. There is, he estimates, at least 1,500 pounds of scientific gear on board, including a big freezer for collecting hailstones, auxiliary batteries, satellite receivers, and anywhere from six to ten HITPR probes, stowed in slide-out drawers like corpses at the morgue. The vehicle is a roving Wi-Fi hot spot, with four or five computers with Internet access running at all times.

A laptop between the two front seats is running a program called Mobile Threat Net, which bounces NEXRAD radar data off XM Satellite Radio's two orbiting satellites (nicknamed "Rock" and "Roll"), creating detailed pictures of storm activity, lightning strikes, fronts, dew points, and other meteorological data. A second laptop is running a program called Street Atlas USA, made by the mapmaker DeLorme, which shows the grid of highways, streets, and rural dirt roads all across the country—an absolutely critical bit of data for a storm chaser. The roof is bristling with antennas and outfitted with what looks like a huge cargo net suspended from a steel rack, designed to capture hailstones for study—and only the very largest ones will do.

On the front hood is a gridded panel of experimental composite material Tim is testing for an unnamed aerospace company. He has fastened a piece of the experimental material to a shiny aluminum plate attached to piezoelectric sensors that measure the high-frequency impacts caused by hail. He's also mounted a high-speed camera on the dashboard, facing out, to capture the impact events on video. Other impact events are more obvious: Though this particular truck is less than two years old, its sides are bashed full of fist-size holes, as if it had survived an encounter with the Incredible Hulk.

Stowed in various compartments here and there is all manner of emergency equipment, such as rolls of heavy rope, tools, a heavy-duty winch, and not one but two spare tires. There are even a few old-fashioned paper maps.

"If the laptop gives off the blue screen of death, we just revert to paper maps," Tim says casually, although seeing the blue screen of death—that is, suddenly losing Internet access and becoming lost in the perilous confusion of a tornado touchdown—would be enough to give the ordinary mortal a heart attack.

GLOSSARY

Dew point The point at which the temperature is cool enough for moisture to condense

Meteorologist A person who studies the science dealing with the atmosphere and its associated processes

LOOKING FOR TROUBLE

Tim (whose last name, pronounced Sa-MAR-as, is a Greek-Albanian name meaning "saddler") is driving north on South Dakota State Route 63, toward the Cheyenne River Indian Reservation, with four other vehicles trailing along behind like a two-bit circus. Since all but one of these vehicles is bristling with strange-looking equipment mounted on the roof, and because they move across the mostly empty landscape in close proximity to one another, like a multicelled organism, it is apparent—especially to the tornado-savvy locals—that this is a storm-chasing team. The group's arrival in small prairie towns is heralded by a flurry of alarm and a jangling of telephones.

"Storm chasers are in town! Did you see anything on the news? Is there a tornado coming?"

South Dakota 63 is the sort of road tourists rarely frequent—wide-open grassland planted mostly in wheat, corn, and soybeans, wending its way through little farming towns with names like Eagle Butte, Whitehorse, and Hayes. The first thing that appears on the horizon as you approach each town is the local grain elevator—the skyscraper of the prairie. The roads are flat and straight as plumb lines, marked by telephone poles marching toward the rim of the world like an infinity of crucifixes. As the day heats up, the hot asphalt roads—oil roads, as the locals call them—become reflective as water, so that approaching cars seem to be swimming across the surface of a prairie river.

It's late morning, a hot, blue day on the Great Plains. There are not any piling, sharp-edged, cauliflower-like cumulonimbus clouds yet—but that's good, Tim says. If the sky is clear, the earth heats up maximally, and by midafternoon we'll start seeing the immense boiling cloud masses that could turn into the supercharged thunderstorms that meteorologists call supercells. Supercells are largely a phenomenon of the late afternoon. And supercells are what—every so often—spawn tornadoes.

"If we wound up driving smack-dab into the path of a tornado, or at least very close, that would be a great day," Tim says. "That's what we're here for." That, in many ways, is what Tim Samaras lives for. His mission out here is to get close enough to extract secrets from storms, in order to understand how they work and thus protect people who don't deserve to be killed by them.

"I remember when I was a kid, growing up in Denver, my mom forced me to play softball with the other kids, but I hated it—absolutely hated it," he is saying, making an effort to explain to a visitor who he is, how he got here, and why he has always been on this contrary path.

"The coach knew I had zero interest in softball, so he put me way out in the outfield. I couldn't stop staring at the sky, and all the cool storms rolling in off the Rockies. I would stand out there hoping one of them would develop into a big thunderstorm and the game would be canceled. I'd ignore the game completely—pop flies would come my way and they'd just hit the ground and roll. Both benches would be yelling at me, including all the parents. I would be completely tuned out. Finally, after one particular incident like that, the coach told my mom not to bring me back to play softball."

From an early age, Tim Samaras was not a kid who played well with others. Even when he was six or seven years old, he liked nothing better than to sit in his bedroom, all alone, taking apart record players and radios to see how they worked. He assumed the role of "mad scientist," rarely emerging from his bedroom Mister Fix-it shop and rarely playing with other kids.

"Funky, weird, dirty kid's stuff—anything electronic—that's what I did," he says. "That's what turned me on."

There were only two things in this world that really interested him—anything electronic, and anything having to do with tornadoes, lightning, and the great storms of the High Plains. When he first saw the tornado scene in *The Wizard of Oz,* it absolutely electrified him. The whole rest of the movie, the Munchkins, the Wicked Witch, Dorothy, Oz: no interest whatsoever. He suddenly woke up to the fact that the place where he lived had some of the most fabulous weather on Earth, most particularly tornadoes, Earth's most extreme form of extreme weather.

At night, he'd ask his parents to open the drapes so he could watch the lightning. Once, when he was in second grade, he noticed a horrendous rainstorm moving in outside the classroom window; it got so dark that a streetlight came on. He stood up and told the class he thought there might be a tornado coming, but the ancient crone who was his teacher told him to shut up and sit down. His meteorological prophecies went unheeded and unappreciated.

"That did turn out to be a big storm, although not a tornado," Tim says ruefully. "I was not that teacher's favorite pupil. In fact, she was the one who sent a note home to my mother saying she thought I had a hearing problem because I had the unique ability to watch the weather so intensely that I just tuned out the entire world. I'd be off in my own zone in the back of class, staring out the window at the weather. I actually got taken to a hearing specialist who told my mom, 'Your son's hearing is exceptional. It's just that he isn't listening.' "

At the age of nine, he saw his first funnel cloud near his house in Denver, and chased it into a neighbor's backyard, climbing up on a swing set for a better look. Up there on the swing, a skinny, trespassing nine-year-old meteorologist and electronics prodigy, nobody was telling him to shut up, sit down, catch the ball, or pay attention to something he was not interested in. He was just staring up, enthralled, at what was unfolding in the heavens.

He was at home. He belonged. He had found his place in the universe.

Tim's father had a job as a hobby merchandise distributor—he sold toy trains and airplanes to hobby shops. He used to love to fly model airplanes in his spare time; sometimes he'd get up at 3 a.m. to build them. Tim remembers waking up to the smell of airplane glue. The apple, as they say, never falls far from the tree. But his dad had a genuine understanding of his son—and a kind of surprised respect, too. His father had always wanted to be a ham radio operator but was unable to pass the complex code test required to get his operator's license. When Tim was 12 years old, he found his dad's discarded ham radio manuals, studied up, and aced the test. At that point, his dad realized that there was a chance this kid might someday be able to make a living at all this.

His dad put an ad in the local paper: "Boy wants old radios and TVs." Old radios and TVs started pouring into the house. Tim tore the electronic guts out of all of them to see how they worked. TVs were lots of fun because they produced high voltage—he had high-voltage wires crisscrossing the room like crime scene tape. One time he took a TV set, disconnected the picture tube, hooked up wires in their place, and then trailed them out his bedroom window, where he watched in delight as 30,000 volts produced a glowing blue-white "corona discharge." Cool. You could almost *see* death. (That amount of voltage, Tim says, may not actually kill you—but it sure would "wake you up.")

His mother, having given up on softball, was tolerant and understanding. She learned to trust that he knew how to handle himself, because he never seemed to get hurt.

"She basically just said, 'Keep the door closed and don't make too much noise in there.' "

When he was 13 or 14 years old, he built a ham radio transmitter out of an old World War II transmitter with giant vacuum tubes. He sat up late at night, talking to people all over the world. Who cared if he was a lousy outfielder? At night, in his bedroom command center, he was the king of the world, riding the airwaves all across the globe.

But he also loved to listen to Mother Nature herself, in the form of static crashes generated by lightning, which buzzed and howled across the radio waves, bouncing off the ionosphere on the edge of space. He loved to run science experiments, rigging up outdoor antennas to see how much electrical charge he could get while a thunderstorm passed overhead. Often the storm would generate so much electrical charge that he could light up a lightbulb. He was like a 20th-century Ben Franklin (though he was still too young to grow a mustache).

When he was 15, he started sending out Morse code, but the system was so primitive it caused the neighbors' TV sets to go haywire.

"When a neighbor kid started interfering with people's *Jeopardy* time, things got serious," Tim says. "My transmitter was also loud—it created a highly annoying, full-volume buzzing sound. I'd be sending out Morse code messages to somebody in Budapest or wherever, and the buzzing would be so noisy that the Spanish neighbor next door, with her daughter practicing her organ, would start screaming obscenities out the window. Not good."

When he turned 16 and got his driver's license, his dad went in halfsies with him to buy a car—a 1967 Ford Fairlane four-door. The first thing he did was to put a ham radio in the car, so he could talk around the world as he drove to school.

"Needless to say, all the high school girls steered way clear of me," he says. As if to underscore the point, he adds: "Yep, I was that guy who enjoyed running the 16-millimeter film projector at school."

He got a job working part-time in a two-way radio repair shop after school, fixing CB radios, ham radios, and business band equipment for taxi companies. He wound up working a 40-hour week making $2.50 or $3 an hour, and by the time he graduated from high school he had three people working for him.

"Between school and my job, when bad weather was brewing, I'd stop in a parking lot to watch the big prairie storms come rolling in during the afternoon. I especially loved to go up to a favorite 'parking' spot near a rock formation called Red Rocks. Up there, you could look down on the whole Denver area—and you could look up at the magnificent summer thunderstorms rolling by overhead."

Other kids may have used their cars to park with girls; Tim Samaras used his to park for tornadoes.

"Whenever a tornado watch was issued for the Denver area, I'd drive to the edge of town to watch. I had no real clue how to forecast or where to go. I just 'followed the big black clouds.' I started reporting what I saw on the volunteer spotter network called Skywarn, which feeds severe weather information directly to the National Weather Service."

What goes unmentioned in this account of his young adult years is one of the things that's most remarkable: He never went to college. Tim Samaras went on to have a brilliant and successful career as an engineer, inventor, and one of the premier tornado researchers in the world, but he never bothered to sit through freshman chemistry or meteorology 101.

When he was 19 or 20 years old, he finally started coming out of his shell, making a valiant attempt to become more of a social being. He began making friends with women, for the first time. And when he was 24, he finally met a woman he wanted to marry.

At the time, he was working at the Denver Research Institute, part of the University of Denver. He was working in the Applied Mechanics Division, on an oil shale project, and he'd come over to the Chemistry Division to borrow a thermometer. That's where he had a chance meeting with an attractive young secretary, also 24, named Kathy Videtich. Not too long after that first meeting, a couple of her friends were going out to celebrate her new job at a downtown law firm and her last day at DRI. It just so happened that Tim knew one of Kathy's friends who worked at DRI and she invited him to the evening celebration.

So that's how Tim and Kathy met.

"He was unlike any guy I'd ever come across," Kathy recalls. "What struck me when I first met him was that he always had things to do and places to go. He seemed to have life figured out. Decisions were very easy for him."

One other thing she liked about this man: "He was just very honest and down-to-earth; you never had to guess where he stood about something. And he got on very well with his family."

He sheepishly admitted to her that he had a house that was almost completely crammed with disemboweled radios and TVs and every other imaginable kind of electronic junk. For a long time he wouldn't even show it to her. "You wouldn't believe me," he told her.

The two of them were able to talk about anything, and soon they started talking about their dreams of what they were looking for in life. Tim, always on the go, didn't waste time. They met in January 1981, were engaged shortly after that, and married in December of that year. December was the month his parents were married, and Tim figured that would be good luck for them.

In 2008 they celebrated their 27th year of marriage. They have three children: Amy, 25; Jennifer, 24; and Paul, who is 20 and—not surprisingly—very interested in bad weather.

"When storm clouds roll in, Paul does what his dad does: He goes up to the top of the street and starts checking out the skies," Kathy says.

Tim also has a son, Matt Winter, 30, who lives in Des Moines, Iowa, and like his father and brother is fascinated by weather.

Kathy has come to terms with the fact that she's married to someone who is extremely passionate about the weather. In the years of their marriage, she says she has never been out storm chasing (though she is an avid watcher of the Weather Channel at home). One day, she says, she'd like to go storm chasing just for the experience.

For his part, Tim says simply: "I've got the greatest wife in the world. She puts up with all my stuff, and that's no small thing."

Earlier on this particular day, in June 2007, a group of chasers who had rendezvoused at the Super 8 motel in North Platte, Nebraska, had driven up to Murdo, South Dakota, to hook up with Tim's group. Now there are five vehicles, and a total of 11 people in the storm-chasing team. There is Tim, the team leader; meteorologist and expert severe storm tracker Carl Young; veteran storm chaser and newly minted Ph.D. Matt Biddle (who was a consultant on the movie *Twister*); senior scientists Bruce Lee and Cathy Finley, the meteorologists; four of their meteorology students from the University of Iowa; Carsten Peter, the *Geographic* photographer; and me.

But there are also the invisible, ghostly hordes that trail out behind this caravan like streamers of smoke. There are all the people from around the world who, for whatever reason, have become fascinated by tornadoes. Tim gets a couple dozen e-mails a week from these people, like the one from a French girl who asked, coquettishly: "If I come to America, will you take me storm chasing?" There are the thousands of schoolchildren who have attended his lectures, in which he shows slides of tornadoes and never fails to talk to them about the importance of pursuing your dreams. For Tim, it is a dream that has already made great contributions to science and humanity—and may even save the life of one of those children.

"These kids get really, *really* excited," he says. And there are the millions of people who have seen or read about him in *National Geographic* magazine, or on the National Geographic Channel on television.

Tim Samaras is a sort of pied piper of tornadoes, with an untold number of people reading his field logs to follow along. He is, one might also say, living evidence that by following your dreams, you can wind up somewhere wonderful.

Which is where he's trying to get to right now. "It's *extremely* difficult to put things in front of a tornado," Tim is explaining as the caravan heads west across the hard, flat country that Lewis and Clark concluded was more or less useless to the future of the United States. It's a land of grass, wide-open spaces, and a sky so big that occasionally it will reach down and eat you. The primary purpose of this mission is to go find a tornado and lay down probes in front of it to gather the sort of down-to-earth data that will help science understand it. For instance, Tim says, "by getting a temperature reading from inside a tornado, we think we can tell how long it will last. That could be very useful."

The danger and difficulty of doing this is perhaps akin to putting down a cherry pie in front of a charging bull, hoping it will step in it—except that the bull could be anywhere within 1,500 miles of here. It is, in fact, so dangerous and so difficult that until recently there were plenty of experts—degreed meteorologists, big-name scientists with prestigious publications—who said it could not be done. That's why, until the summer of 2002, nobody had ever succeeded in deploying a probe into the center of the cyclone.

Many attempts had been made to do it. In fact, back in 1986, Tim was electrified when he watched a PBS *NOVA* show called "Tornado!" that featured a scientist attempting to put down a probe in front of a tornado. The probe, dubbed TOTO (after Dorothy's dog, of course) looked like a 55-gallon drum and was loaded with meteorological instruments that would measure barometric pressure, temperature, wind speed, and direction.

Wow, Tim remembers thinking at the time. "There really *are* people out there that really chase storms!"

TOTO (which stands for Totable Tornado Observatory) had been developed by the National Severe Storms Laboratory, in partnership with the University of Oklahoma. But getting the probe smack in the center of the cyclone proved so difficult that after years of trying, the NSSL essentially gave up.

But TOTO had one big thing going for it: The idea was highly cinematic, which is why, in the hit 1996 movie *Twister,* Helen Hunt and Bill Paxton (who play tornado researchers) keep trying to put a probe called Dorothy directly in front of an F5 twister. Dorothy looks suspiciously like NSSL's 55-gallon drum, which is packed with hundreds of tiny golf ball–size probes. When Dorothy is taken up by a tornado, the drum is supposed to break open like a piñata and disperse the probes up into the funnel cloud, producing an electronic image of the entire deep structure of the storm. Or so the movie goes.

Though most other researchers gave up on the probe-deployment idea in favor of other ideas like "mobile mesonets"—weather stations mounted on the roofs of cars and trucks, such as the truck-mounted Doppler radar stations known as DOWs (Dopplers on Wheels) and other systems—Tim Samaras persisted.

In his job at Applied Research Associates, Inc., scientists and engineers often challenged him to difficult tasks, crafting new electronic gadgets that did strange things. That was his specialty, honed in his bedroom as a kid: making electronic gadgets that did strange things. He thrived on the incredible challenge and difficulty of doing things like this. And he seemed invariably to come up with answers and measurements that were extremely difficult to achieve.

"Looks like there's some new convections firing up around Rapid City," Tim is saying over the two-way radio. "Maybe there's some juicy fronts we can work with over there."

After a brief consultation via two-way radio among the best meteorological minds on the team, the five vehicles in the chase group make a midcourse correction and go booming west on 212 toward Rapid City (or Rapid, as the locals say, rapidly). Every ten or fifteen miles there's another sign for Wall Drug, the West's kitschiest tourist attraction, in Wall, South Dakota. At South Dakota's National Grasslands Museum, a display explains that after the Homestead Act of 1862, "a massive advertising campaign described the Plains as the greatest opportunity since the Garden of Eden." (It had previously been known as the Great American Desert, a pitch certain not to attract a crowd.) The peak period was 1875 through 1890. Now, more than a century later, huge areas of this country are probably not much more heavily populated than they had been before the ad-fueled land rush. The emptiness of this country is so enormous it becomes a presence all its own.

"All this emptiness and flatness makes this a perfect place to see tornadoes," Tim says. "In places like, say, Wisconsin or Arkansas—where we've also chased tornadoes—there is too much tree cover and too many hills to be able to see very far."

To Carsten, who has traveled widely, the landscape is reminiscent of Russia—the vast, wide-open spaces, the large-scale agriculture, the little run-down, somewhat melancholy towns.

Among the various wireless programs Tim has running on the laptops in the geekmobile, Mobile Threat Net seems to be getting a good deal of use today. Developed by a Huntsville, Alabama–based company called Baron Services, it is essentially a boiled-down version of the kind of high-end technology used by professional TV weather forecasters. On a laptop computer screen, the user can get an amazing array of detailed weather data, beamed in from 158 NEXRAD radar sites, which update every five minutes. The information is delivered (wirelessly, of course) through the powerful satellites of XM Satellite Radio, and unlike some similar systems, it still works when the weather gets really, really bad. It allows the user to zoom in on weather maps right down to ground level, showing reasonably detailed radar images, smooth and clear as TV and not so much like blurry, blocky pixel images. It gives you detailed pictures of severe thunderstorm cells and storm tracks, precipitation types, tornado warnings, and reports from NOAA's Storm Prediction Center. It gives real-time nationwide lightning information, as it occurs. It shows winds at Earth's surface and up to 42,000 feet aloft. It depicts rotating winds or wind shear as little spinning disks—the jackpot in the casino of storm chasing. And, in the midst of all this activity, it tracks your own vehicle's location by means of GPS.

GLOSSARY

Mobile mesonets Moving observation stations that study mesoscale weather and its associated phenomena

On the other hand, though one would think that these new technologies are so precise they make storm chasing a snap, it's not true. There may be a wealth of new data, but it is difficult to interpret, particularly in light of the fact that you have to deduce what the weather is *going* to do long before it does it. Otherwise, you'll never get there in time. And the system does have its shortcomings. Some chasers complain that the radar data, though bright and compelling, "smooth out" multiple radar images and are too "cartoonish" and lacking in detail. And then there are the time lags between updates, which may be five minutes long but sometimes much longer. In the muddy chaos of a storm, information that is five minutes old can be about as up to date as the Dead Sea Scrolls.

Because of this, Tim says, "you should never make life-or-death decisions based on radar information."

(The other gizmo that has Tim's attention today is the hail-impact platform mounted on the hood of the truck. Today looks ripe for hailstones, especially the killer kind over two inches in diameter, and that's exactly what a certain aerospace company wants to find out more about.

"What they want to know is what happens to the skin of their new aircraft if the plane is parked on the tarmac on a nice spring day in Dallas when a huge hailstorm rolls in," Tim explains. "How I'm planning to figure this out is that I have essentially turned my whole vehicle into an in situ hail sensor. On the hood of the truck is a half-inch gridded plate, five foot by three foot, made of their new aircraft material. Linked to the plate is an accelerometer, which basically measures the shock wave of the impact. The accelerometer is also linked to a trigger on a Phantom high-speed camera, which is mounted on the dashboard and pointing out through the windshield at the plate. The accelerometer also triggers high-intensity lights. Hailstones trip the accelerometer and then the camera and lights, which then records impact information plus pre- and postimpact information."

The setup is pure Samaras wizardry: deft, ingenious and automated, with as little user interaction as possible. The camera can expose as many as 15,000 images per second, compared with 30 to 60 frames per second for a typical camcorder. However, Tim explains, since high-speed photography is always a compromise between speed and quality, he usually shoots much more slowly, at around 1,000 frames per second.)

Despite all the gee-whiz equipment and the four wireless computers sucking in great masses of data out of the air, Tim explains that it's still enormously difficult to predict tornadoes, much less get there while they still exist. When you make a space-time comparison of the life span of a typical tornado (most last less than ten minutes) with the extent of the domain where they are most likely to occur—from southern Canada south to the Rio Grande, west to the Rockies and east to Indiana—you begin to understand how challenging it is to get to precisely the right place at precisely the right time.

Often Tim's team will hear of a tornado touchdown that is simply too far away to be reached. Or he'll be hot on the trail of one, but it is simply moving too fast, or the roads in that area are too bad, to follow. But the challenge of finding a touchdown also adds to the richness of those rare moments when you are actually there to witness one of the greatest spectacles in nature. That's why Tim and the other storm chasers tend to talk in a kind of shorthand of remembered moments—"Childress, Texas," "Storm Lake, Iowa," "Manchester, South Dakota." Or, perhaps, wherever they are later this afternoon.

In some ways what all these storm chasers are actually seeking are those moments of supercharged aliveness, an ecstatic contact with the now so potent that its memory will be branded into the brainpan forever. And the fact that this rapturous contact is not just an idle dance with death, a suicidal thrill, but a deeply serious search for scientific understanding, imparts a profound pleasure to the chase.

Storm chasing is also a way of jacking life over into a realm of truly magnificent problems. Human life is essentially an endless series of problems—problems so numerous, so relentless, and so time-consuming that we tend to *become* our problems. And because most of our problems are boring and trivial, our lives tend to become boring and trivial. But out here, instead of worrying about whether the utility bill got paid or Johnny got to soccer, you are worrying about whether you will live or die, and if you die, whether it's to be by lightning, hail, or being taken up to the gates of glory in a whirlwind.

To storm chasers, *those* are problems worth having, and a life worth living.

Now, to the west, we can see huge, flat-bellied cumulus clouds marching across the sky like flotillas of battleships. The sky begins to darken and fills with drama, like organ music. And Threat Net begins to report something interesting brewing still farther to the west, over the Wyoming border toward the Thunder Basin National Grassland. Something exciting is afoot in the atmosphere.

"Tower going up at our one o'clock!" Matt radioes to Tim.

"Yep—I see it!" Tim radios back, craning out the windshield to see past the gizmo-crammed cockpit.

The thunderheads spread fat, fast-moving cloud shadows across the green-golden earth. Invisible to the eye, some of these budding storms are also drawing along beneath them magnetic "shadows" bearing a positive charge, in opposition to the cloud belly's negative charge—an electric differential that will become dramatically visible as lightning if the storm fires up.

Matt Biddle is getting excited.

"It's cookin'! It's lookin' good!"

Matt, who walks painfully with crutches due to a congenital medical condition compounded by a recent stroke, is nevertheless a crack high-speed storm chaser. He's like Ahab, transformed by the chase for the great white whale.

"I come alive when it storms!" he howls.

"There's an incredibly juicy air mass lying over the west-southwest, nice dew point depression, 78/70, with winds from the east-southeast at 10 gusting to 20," Tim is saying over the two-way radio. "I think we should just go toward Rapid City and kind of caress it."

The caravan of chasers, with Tim's geekmobile in the lead, makes another course correction and heads almost due west. It's as if some fearsome, gigantic weapon high up in the atmosphere is wheeling around and directing its aim toward Rapid City—though most of the people who live there probably do not yet know that they are in its crosshairs.

Along the riverbeds, you can see cottonwood trees releasing their silvery snow into the wind, and fields of golden wheat. There are huge, round hay bales laying in the fields that, when freshly cut, shine like a girl's hair but gradually grow duller and browner, more like loaves of brown bread, as they age. We see the occasional ring-necked pheasant, once or twice a roadrunner, and a couple of scissor-tailed flycatchers teetering on telephone lines. The roadkill gives other clues about where we are: dead armadillos along the road, and coyotes, sprawled in smashed sleep.

In the early afternoon we get the "Day One" report from the Storm Prediction Center in Norman, Oklahoma. The chance of a tornado in the area where we're headed has dropped from 15 percent to 5 percent, but Tim is convinced the SPC is wrong. We're still heading west toward Rapid City.

Tim likes our chances today. He can't understand why the SPC is "ignoring" South Dakota. He thinks it's a "bull's-eye" over Rapid. We get there by mid-afternoon, a baking, blue Dakota summer day. There does not seem to be any sign of a tornado, or even a pussycat of a thundershower. We stop at a sandwich shop and run into a guy in line who says he lived through the 1957 tornado in Fargo, North Dakota. That storm was made famous by Ted Fujita (see Chapter 10), who became known as Mister Tornado—a brilliant researcher and professor of geophysical sciences at the University of Chicago who made key discoveries about tornadoes by studying them.

We step outside the sandwich shop and just kind of wait for the weather, while around us the storm system that produced the initial tornado warning rapidly falls apart. Tim calls it "caca." The weather is good. And that's bad.

Around five o'clock we stop for fuel near Spearfish, South Dakota, but it's Sunday and the town's lone gas station is closed. Around us, immense thunderheads tower up to the edge of the stratosphere. Aldous Huxley once described such muscular atmospheric creations as "vague torsos of fabulous athletes."

A kid with a mop of curly hair and great dark eyes shows up on a bicycle. He's eating us up with his eyes.

"What are you guys doing?"

"We're looking for tornadoes."

"You're storm chasers? It's an awesome profession!"

He tells us his name is Donovan, and that his grandfather was a reporter, that he photographed tornadoes in Nebraska and once even got picked up by an F2. He photographed Charles Starkweather's murder spree in Nebraska and Wyoming back in 1958. Donovan tells us he has a severe-storm watch on his iGoogle page, and that he once saw a TV show about guys tracking tornadoes on motorcycles. That was on *Extreme Careers.* He's 17, casting about, growing up in a tiny town in the Dakotas, and tornadoes seem to have seized his imagination like a holy revelation. He peers into the car, intently watching Tim, Carl, and the others studying Threat Net on the computer screen. His big, dark eyes seem to devour everything.

And then, all around us, the storm system just seems to fizzle out and blue sky appears. Storm chasing, for the most part, is lots of driving and lots of disappointment. We get out of the cars in a field near a cemetery filled with Norwegian and Scandinavian names—Torvoldson, Knutson. There's a moment of emotional deflation as the system seems to collapse and then go crawling across the prairie sky to the northeast. Around us there's a gorgeous prairie sunset. Tim is quiet, sprawled across the hail impact platform, staring up into the heavens. It's nearing the end of June, and thus the end of the storm-chasing season. This may be the last system he'll get a chance to chase this year.

Around us huge breakers of wind go undulating through the grass—the wind and the grass roll in great golden whitecaps for a thousand miles across these plains. We can hear "tornado birds" in the grass. A guidebook describes their call as a "rich flute-like jumble of gurgling notes, usually descending the scale," but there is also a whooping sound, like some exotic bird of the tropical rain forest. Barn swallows do delirious barrel rolls in the blue air, windsurfing. And the red-winged blackbirds, happy that it's been such a wet summer and there is standing water everywhere, teeter on cattails, stridently announcing to the world their single observation: *"Conk-a-ree!"*

One of the meteorology students is wearing an American Meteorological Society T-shirt whose back is completely covered with algebraic equations. Underneath them, it says, "It's So Simple!" He leans back against one of the cars and shouts up at the clear, empty sky: *"Stupid atmosphere! Why does it always have to obey the laws of physics?"*

It's a weather-geek joke—but it sums up the way most storm chases end, in some rumpled field like this, beneath the ragged, fleeing remnants of a dissipating storm.

4: INTO THE CENTER OF THE CYCLONE

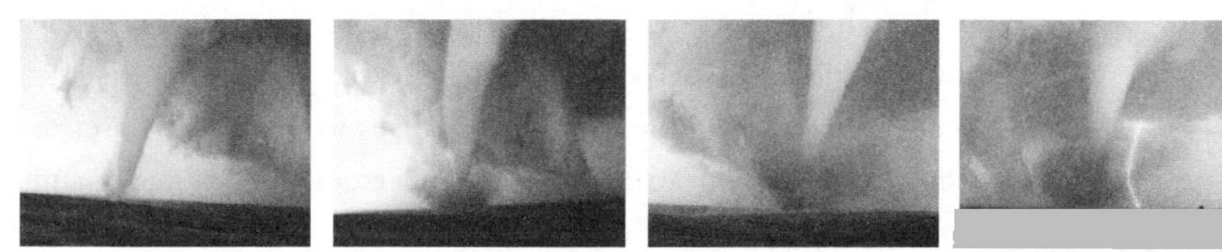

BACK IN THE 1930S, A BRILLIANT ESTONIAN TORNADO RESEARCHER NAMED JOHANNES Peter Letzmann wrote a letter to the U.S. Weather Bureau. Knowing that the United States was home to more tornadoes than any other country in the world, Letzmann wanted to propose that the Americans set up some kind of organized program for watching tornadoes and waterspouts (tornado-like vortices that occur over water). The Weather Bureau's official response sounded not so much like an agency tasked with boldly going where no man had gone before, but instead more like an irritable old lady afraid of getting her feet wet:

> **[Regarding] a tentative schedule for investigation of waterspouts and tornadoes proposed by Prof. Letzmann . . . we are convinced that it can never be fulfilled in this country. We have on the average nearly 150 tornadoes a year, scattered over a vast area, and only very occasionally at many year intervals is there opportunity for a meteorologist to actually observe an occurrence. . . . In all the years of record, a tornado has never passed over a set of recording instruments, and in case one should, they would be demolished or carried away. People who have never witnessed an American tornado have no conception of its horror,**

and therefore it would be utterly impossible to make observations other than of a very general character. . . .

Fortunately, human knowledge of Earth and its atmosphere is not in the hands of people like this frightened, now forgotten functionary of the Weather Bureau (who trembled at the thought of 150 tornadoes a year, though the real number is closer to 1,300). Instead, it is in the hands of people like Tim Samaras—daring explorers, filled with passion and ingenuity, who have risked their lives to push ever closer to the whirlwind's core in an attempt to understand its mysteries and protect the people in its path.

Tim's efforts to place an instrument-laden probe directly in the center of the cyclone was a daunting—many said impossible—task. The core is so difficult to access, so short-lived, and so utterly hostile not only to human life but to almost any kind of measuring instrument—one thing the bureau was right about—that to this day it remains a kind of "black hole of meteorology," in the words of scientist Anton Seimon.

In his in-depth look at the long and fascinating history of attempts to penetrate a tornado's core with instrument probes, Samaras points out that since the late 1970s, various researchers have built various devices and gotten them somewhere close to a tornado, with varying degrees of success (or, more often, failure). This motley parade of drones, robots, disks, and 'droids weighed anywhere from 400 pounds all the way down to 4 pounds. Their purposes were also varied, from measuring wind speed, temperature, and humidity to electrical activity, sound waves, and seismic vibrations in and around tornadoes.

One of the most revealing and valuable measurements from the tornadic core is barometric pressure. As in hurricanes, barometric pressure is the most dependable indicator of a tornado's intensity; the lower the barometric pressure, the more intense the storm.

A mercury barometer has a glass tube filled with mercury, closed at one end, with an open reservoir at the base. High atmospheric pressure exerts force on the mercury in the reservoir, which raises the mercury level in the tube. Low pressure has the opposite effect, allowing the mercury level to drop. Low barometric pressure is associated with precipitation, storms, and tornadoes because when a low-pressure area moves in, air expands and rises. When air rises, it weighs less than air that is sinking, so barometric pressure falls. Rising air also cools, then condenses, which leads to clouds, precipitation, and storms. That's why when a big storm complex is headed your way, it is usually preceded by falling barometric pressure, hours or even days before the storm itself arrives.

Barometric pressure can be denoted in inches or in millibars (a metric measurement). It may be a bit surprising how little barometric pressure rises or falls: Normal barometric pressure at sea level is about 30 inches, and the lowest barometric pressure ever recorded was 25.69 inches (in Typhoon Tip, a legendary hurricane in the western Pacific). A millibar refers to one one-thousandth of the standard air pressure at sea level, which is about 1,000 millibars (1,013.25, to be precise). The 100-millibar pressure drop that Tim Samaras measured in the Manchester tornado was phenomenal.

Home barometers were the first "probes" ever to penetrate a tornado—although this occurred not on purpose but by accident. One of the first of these accidental measurements occurred in 1896, when a tornado passed over St. Louis and a level-headed citizen was able to observe an 82-millibar drop in pressure on a home barometer. In 1904, a tornado hit Minneapolis and another citizen reported having observed a—literally unbelievable—192-millibar pressure drop, also on a home barometer. (This reading is considered highly questionable, however, because it was taken "under considerable duress," and because it is far higher than any other measurements ever recorded. Of 16 other such accidental measurements, the average is 19.4 millibars.)

But fantastic barometric pressure drops aren't the only phenomena that have beckoned to scientists from inside the core. In the early 1970s, a physicist from the University of Mississippi, Roy Arnold, became fascinated by the *sound* of tornadoes and began attempting to deploy audio recording instruments within or close to the path of tornadoes. The goal of this program, known as Sound Chase, was to determine if a tornado produced an identifiable sound wave that could be used as a "signature" for identifying—and anticipating—tornadoes. What was especially intriguing was the possibility that tornadoes might produce very-low-frequency sounds, which (unlike higher frequency, audible sounds) can be detected hundreds or even thousands of miles from their source. If it were possible to identify some low-frequency sound signature, it might be possible to trigger a "tornado alarm" a great distance from the storm. That would be a huge improvement over relying on human storm spotters or radar, neither of which can perfectly or reliably predict tornadoes.

GLOSSARY

Waterspout A small, relatively weak tornado that occurs over water, most common in tropical or subtropical waters

Seismic Relating to the vibrations or movements caused by earthquakes and other phenomena

Storm cell The convection in the form of updrafts or downdrafts, seen in towering cumulus clouds; a typical thunderstorm is made up of a number of cells.

So many tornado survivors have described the sound of an onrushing twister as a deafening roar akin to a jet engine or a freight train that it's almost a day-after cliché from every newspaper tornado story you've ever read. But veteran storm chaser Howard Bluestein, a meteorologist at the University of Oklahoma, says in his book *Tornado Alley* that "in all my years of storm chasing, I have never heard the freight-train roar many say they have heard in a tornado. . . . I suspect that one has to be much closer, dangerously close, to a tornado to hear the fabled roar."

David Hoadley, who began chasing storms back in the early 1950s, made some interesting observations about the sound of tornadoes in an early issue of *Stormtrack,* one of the first storm-chasing publications.

Although he had seen and photographed more than 60 tornadoes and funnels, Hoadley said, he heard the famous roar only one time, and that was at a moment when no tornado funnel was visible. He was chasing a storm cell near Caldwell, Kansas, in 1968 and stopped to watch an immense rotating cloud (whose tops on Wichita radar were reported at about 67,000 feet) when he heard a "distinct, loud and continuous roar (as hundreds of jet planes high overhead)." That's when he noticed something odd. Although he assumed that the sound was emanating from the rotating cloud, when he slowly turned his head from side to side to pinpoint the source of the sound, he was surprised to discover that it seemed to be omnidirectional—coming from every direction at once. He conjectured that somehow the overcast sky might have created some kind of echo chamber effect. It was an unproven idea, but one that suggests how much mystery still surrounds the acoustics of tornadoes.

One characteristic of the sound of a jet engine or a freight train is that, in addition to being enormously loud, it's also accompanied by a deep, earthshaking vibration—and perhaps a low-frequency sound wave. Sound Chase, one of the earliest attempts to place probes in front of tornadoes, succeeded in asking such tantalizing questions but was not able to answer them fully.

"A tornado is a very long, whirling tube of air, an enormous acoustical instrument, with a hollow core and debris-filled cone or cylinder," observes Tom Grazulis in his 2001 book *The Tornado.* "No one has fully explored the sound-generating properties of such an object."

In 1995, engineer Frank Tatom, of Huntsville, Alabama, also became intrigued by the deep vibrations produced by tornadoes and designed a small two-part probe called a snail to test whether twisters might produce some specific seismic signal, like an earthquake. Based on several eyewitness accounts and a few seismic recordings during tornado touchdowns, Tatom became intrigued by the idea of trying to isolate a seismic signal to be used as an early warning of tornadoes, much like Arnold's low-frequency sound wave.

Tatom and two colleagues reported that there was no evidence of this tornado seismic signal in scientific literature. Nevertheless, they had strong reason to believe it might exist. They had also gathered intriguing eyewitness accounts:

In one case, a service station manager was working on replacing the headlight of a car in front of his service station in Huntsville on November 15, 1989. The vehicle was facing east, with the hood raised and the engine turned off. As he was working, he "began to notice strange vibrations in his feet, and he also noticed that the engine within the car was shaking." Although he could hear no sound and the wind was not blowing, he glanced up and noticed that the sky had turned a peculiar color and he "realized something unusual was about to happen."

He urged the driver of the car, a woman, to run inside the station office. When she refused, he "pulled her from the car and escorted her into the office," where the two of them huddled behind the counter. The office windows were vibrating eerily, though there was still no wind. Less than a minute after feeling the first vibrations, the man heard the roar of the tornado. Moments later the tornado, later rated an F4 (winds between 207 and 260 mph), struck the service station and demolished it. He and the woman were swept out of the building into the middle of the street, terrified but alive.

Excited by stories like this, and by the fact that there really was no way of detecting a tornado on the ground other than by human observation, Tatom envisioned a seismic tornado detector that could alert people to the danger of an oncoming tornado, perhaps even long before it could be seen or heard.

To gather the evidence that such a seismic wave existed, Tatom's "snail" probe consisted of a little rectangular box containing a "geophone," to detect seismic signals, and a separate dome-shaped device containing the recorder and battery. Tatom loaned out this contraption to chasers to deploy in storms. On Memorial Day weekend of 1997, Tim Samaras became the first—and so far the only—person to deploy a snail near a tornado, in this case, a quarter-mile-wide tornado in south-central Kansas. The instrument detected a strong signal as the tornado passed by.

"I think Frank is onto something," Tim says, "But the question is whether there is a seismic signal there that is reliable enough to build a tornado warning system around." For now, at least, Tatom's seismic tornado detector—which would represent a "totally new way of detecting a tornado on the ground," in his words—has yet to be built.

Another kind of tornado probe, this one pulled by an airplane, was inspired by the experience of a young doctoral student at Florida State University, who one day in the late 1960s got a chance to use his brand-new Super 8 movie camera and discovered his life's work in the process. Joe Golden and a friend were taking a pleasure flight over the Florida Keys when they spotted a waterspout a mile or two away, and Golden was able to film it. The excitement he experienced that day expanded into a doctoral thesis and a lifetime of work studying waterspouts (whose vortices are similar to those of tornadoes but generally weaker).

He began flying around and then *into* waterspout vortices. On his early flights, he would rent a plane and then ask the pilot to fly near waterspouts; then he'd throw smoke flares, balloons, or confetti from the plane to reveal the direction of air currents in and around the vortex. Not exactly high tech, but it worked.

During the month of September 1974, the first ever in situ measurement of a vortex took place over the Keys when Golden rigged up a light aircraft with a cone-shaped instrument package (a bit like a traffic cone) to be pulled along behind the plane. During a 12-day period from September 16 to September 27, Golden penetrated 16 waterspouts, getting data about their temperature, humidity, and dynamic and static pressure. This was a genuine breakthrough, and one that showed that it was actually possible to get *inside* the vortex—if not of a tornado, then something very like it.

Another airplane-deployed probe was devised by a colorful tornado researcher named Stirling Colgate, from Los Alamos National Laboratory (and an heir to the toothpaste fortune). Colgate got the idea of trying to fire ten-inch-long, instrument-laden rockets directly into a tornado funnel from an airplane. Weighing less than a pound each, the rockets were loaded with miniaturized weather instruments to measure air pressure, temperature, and various kinds of electrical activity. Colgate had been fascinated by science since he was a boy. But it was only later in life, in his 50s, after he resigned from his job as president of the New Mexico Institute of Mining and Technology, that he turned his attention to a lifelong interest: tornadoes. He was especially interested in the electrical properties of severe storms and tornadoes.

He rigged up a Cessna 210 aircraft with the little rockets, loaded onto the undersides of the wings so that they could be fired remotely from inside the cockpit. Once the rockets were fired, he used radio telemetry to transmit the data from the rockets back to a computer aboard the plane.

One problem with this scheme was that, according to Federal Aviation Administration regulations, the rocket motor was allowed to contain no more than 80 grams of propellant, to meet federal guidelines for a nonlethal weapon. In other words, it was a fairly wimpy rocket. Another problem was that, during spring 1981, he spent more than 120 hours in the air but encountered only a single tornado. Colgate tried numerous times to fire his little rockets into this and other tornadoes, but none appeared to penetrate the tornado funnel (although he did get some data about ionization, electrical charge, and pressure drops).

Colgate started out flying around and into waterspouts in the Florida Keys, like Joe Golden, because that was safer than tornadoes. Then he moved his operations to Norman, Oklahoma, home of the National

Severe Storms Laboratory. For three seasons during the early 1980s, Colgate flew his Cessna 210 laden with tiny rockets out to try to find the beast, like St. George seeking the dragon.

On May 22, 1981, he ran into a tornado 70 miles west of Oklahoma City. He fired four rockets at the funnel and missed every time; the last one fired when he was only a few thousand feet from the black wall of the funnel. (These near misses were captured on a Super 8 movie camera and a 16-millimeter movie camera, one attached to each wing.) The next season, he tangled with a huge tornado over Pampa, Texas. He was nearing the funnel when the airplane hit a powerful downdraft. He was swept down toward the earth like a fleck of dust, desperately struggling with the plane's controls as it neared the ground. According to an account in Keay Davidson's book *Twister,* Colgate later said he got so close to the ground that he could see "the television picture in someone's living room" before righting the craft and regaining the air.

After that fateful flight, Colgate never again tried to fire a rocket at a tornado.

In the early 1980s, a probe wryly dubbed TOTO made its appearance. TOTO stood for Totable Tornado Observatory and also, of course, for Dorothy's little dog, who took an unscheduled trip to Oz via twister. Designed by two scientists at NOAA, Al Bedard and Carl Ramzy, TOTO actually wasn't all that totable, since it weighed about 400 pounds. TOTO looked vaguely like a cross between a hot-water heater and R2D2, the tottering 'droid in *Star Wars.* Roughly the size of a 55-gallon oil drum and encased in a half-inch-thick aluminum cylinder, it had a vertical pipe on top laden with instruments to measure wind speed, wind direction, temperature, air pressure, and corona discharge. (Corona discharge is what causes the crackling sound you hear underneath high-voltage power lines—a side effect of the interaction between an electric field and the surrounding air.)

Tim Samaras holds a probe that calculates tornado wind speeds.

The data were recorded on mechanical impact instruments that captured a data point every second.

The fictionalized version of TOTO, Dorothy—introduced to the world in the 1996 movie *Twister*—would, after making contact with a tornado, break apart like a piñata and disperse hundreds of glowing, Ping-Pong ball–size transmitters up into the heart of the storm. The concept made for good cinematic effects but did not (at least so far) depict anything tornado scientists have actually designed.

TOTO was considerably more down-to-earth, considerably more cumbersome, and not nearly as sexy as Dorothy, but deploying it was every bit as dangerous in real life as it was in the movie. That's why Bedard and Ramzy figured out a way to quickly offload the device from a pickup truck with a ramp and winch system. They claimed to have been able to do this in 12 seconds. One special gadget to speed things up was an automatic turn-on feature in the device. When TOTO lay on its side, it wasn't turned on. Tipped into its upright position, mercury tilt-sensitive switches automatically turned it on, activating all the sensors and recorders inside.

But TOTO, like all its predecessor probes, had its problems. One serious shortcoming was that, when TOTO was tested in a wind tunnel, it was discovered to have a center of gravity that was too high, so the whole thing could tip over (and thus turn itself off) in winds as low as 114 mph. In the face of a mere EF2 tornado (with wind speeds of 113 or higher), poor TOTO would tip over like R2D2, whimpering its sad little machinelike whimper, and then go silent.

Despite many attempts, TOTO never actually got inside a twister. Bedard, Ramzy, and the other NOAA researchers blamed the short-lived nature of tornadoes, bad roads, and the fact that TOTO was cumbersome to deploy. It was decommissioned in 1987 and is now, like the Mercury space capsules, a museum piece of the history of human daring and exploration.

One thing TOTO succeeded in providing, though, was that intangible commodity that changes the world: inspiration. During the one time that scientists succeeded in putting TOTO close to a tornado's path, a film crew from the PBS show *NOVA* happened to be present. That show, called "Tornado!" aired in 1986 and electrified a young engineer from Denver named Tim Samaras, who started thinking about and then designing tornado probes of his own.

Meanwhile, in the late 1980s, other tornado researchers—this time, Fred Brock, Glen Lesins, and Robert Walko from the University of Oklahoma—were also learning from TOTO's problems and trying to design probes better suited to the task of penetrating the core. Their probes, much smaller than TOTO, looked less like R2D2 and more like small, shiny, overturned salad bowls. Which is exactly what they were. You can't call a serious scientific instrument a salad bowl, of course, so they became known as turtles.

The turtle probes were a mere 14 inches in diameter, with a 55-pound lead weight molded around the inside circumference to keep them fastened to the ground (an innovation that foreshadowed Tim's HITPR probes). On May 16, 1986, University of Oklahoma researchers chasing west of Wheeler, Texas, were able to deploy several of the probes near the path of a storm. They captured a gradual drop in atmospheric pressure during the first 30 minutes of the storm's development, which they interpreted to mean that updrafts drawn up into the storm's southern edge were lowering air pressure in the storm's outer flank.

Bill Winn of Langmuir Laboratory, in New Mexico, constructed instruments known as E-turtles to record pressure, temperature, and the electric field of tornado cores. They were fielded as part of a much larger project, called VORTEX, during 1994 and 1995. On June 8, 1995, an E-turtle was deployed on the edge of a large tornado near Allison, Texas, measuring a pressure drop of 50 millibars, one of the highest credible numbers yet achieved.

"That was a fantastic measurement," Tim says. "At the time, a real breakthrough. And Bill's E-turtles really were a predecessor to my probes. But like every other design, they had problems—for instance, the accuracy depended on which way the wind blew."

Tim Samaras brought an unusual collection of skills to the task of designing a tornado probe of his own. He knew a great deal about electronics, instrumentation, computer modeling, and meteorology. As an expert in explosions, he was also familiar with the materials that could withstand phenomenal explosive power, like the "atmospheric explosion" of a tornado. At Applied Research Associates, where he worked as a senior engineer, he had access to wind tunnels, where designs could be tested in something like the actual chaos of a real tornado. He'd been carefully studying the field-testing of the whole crazy parade of probes, 'droids, and disks that had been deployed into tornadoes by other researchers to figure out what worked and what did not. And perhaps most important of all, he had a wealth of practical experience chasing tornadoes.

"If anybody could get a probe in front of a tornado, and design it to survive, it's Tim Samaras," says storm chaser Tony Laubach, whose admiration for Tim seems boundless.

In 1997, Tim developed the first version of his HITPR probes. His design, which was made of quarter-inch-thick steel, looked like a flattened cone about the diameter of an auto tire. It weighed 45 pounds or so and cost anywhere from a couple of thousand dollars up to $15,000, depending on how instrument laden it was.

He began working with that conical shape because, when he ran computer models on hemispherical dome shapes (like the NSSL "turtle" probes), he discovered they had a little bit of lift. It was too easy for tornadic winds to peel them off the ground and turn them into just another fragment of flying debris.

Samaras and his team at Applied Research Associates were also partly inspired by the work of Roy Heiman, an engineer who worked at the aerospace company Lockheed Martin during the early 1970s and 1980s. Heiman had helped design a hardened missile silo for a mobile missile launcher, which would pop up and then move along the ground. He discovered that a triangular shape was most resistant to the phenomenal explosive power of a nuclear attack. Samaras tried a triangular shape that had a circular base—a cone—and then ran the computer code on this shape. He found that it worked. A conical shape, tapered at a 30-degree angle, broke up that lifting action, even when it was aerodynamically tested in a wind tunnel at speeds up to 225 miles an hour. The cool thing about the aerodynamics of the device was that the harder the wind blew, the more the device pushed *down,* rather than lifting *up* (or tipping over and turning off like TOTO). In a tornado it was like a lamprey stuck onto the side of a shark.

Another advantageous thing about the design of the HITPR probe was that it addressed the problem of measuring error. It was actually very difficult to get accurate barometric pressure readings in a tornado because the wind flow over any measuring device, especially if the winds are over 160 mph, is extremely turbulent as it passes over the surface of the probe. But Tim's design adjusted for this problem by compensating for the curve of the air flow while pressure is being measured.

So Tim's "cool orange hat" was a design that was able to get accurate measurements even in extremely high winds; could measure barometric pressure, air temperature, relative humidity, wind speed, and direction near the ground; would not blow away; was easy to transport and relatively inexpensive; and was very difficult for a tornado to destroy.

Once the probe was deployed in the path of an oncoming tornado—that is, hustled to the edge of some run-down road in rural Kansas or Nebraska—Tim still had to activate it using a little switch on its underside. He'd switch it on, then he'd flop it on the ground, jump back in the van, and take off like a bat out of Hades. Meanwhile, behind him, his faithful little tornado robot would be fast at work. The internal data loggers could record 18 channels of information for up to two hours. There was enough onboard memory to deploy the probe dozens of times in succession without having to download any data. Later versions also had six or seven tiny video cameras inside, arranged to produce a 360-degree view through little super-hard Lexan windows—making the whole contraption look even more like a tour bus from Mars.

Though whole chase seasons could go by with nothing but a distant tornado sighting, or none at all, any day that produced an up-close encounter with a twister, but most especially one that produced any kind of scientific data at all—a barometric pressure reading, a revealing bit of video—would be a good day indeed.

The HITPR probes filled in a kind of empty biological niche in the world of tornado exploration. The giant truck-mounted Doppler radar dishes called Doppler on Wheels (pioneered by tornado researcher Josh Wurman) were able to show dazzling detail of tornado cores—but only above 50 or 100 feet off the ground. It was difficult to get good radar pictures that close to the ground because of obstructions like buildings and trees and because of the physics of microwave signals.

Why is getting information about the lowest hundred feet of a tornado important?

"Because that's where we all live," says Tim. "That's the space we occupy on the planet."

Also, he adds, researchers still don't understand the relationship between wind velocities at higher altitude and what's going on close to the ground.

That's why Tim's "dream data set" would be to have a very violent tornado that is penetrated simultaneously by an HITPR probe on the ground; several mobile mesonets at mid-level; and mobile Doppler radar combing the higher elevations. That way, for the first time, it might be possible to see the whole interior architecture of a tornado—all the complex updrafts and downdrafts, the mesocyclonic rotation, the inflow and outflow that feeds the storm and keeps it alive.

Beginning in 2002, Tim and his team began taking these probes out into the field, getting closer and closer to the center of the cyclone. They successfully deployed a probe in an F3 tornado near Pratt, Kansas, on May 7, 2002. The deployment demonstrated that the little orange hat worked like a charm, and that it could gather great data—in this case, showing a 24-millibar pressure drop at the edge of the funnel. A video camera was also placed near (but not inside) the probe, and wind speeds estimated from objects being hurled across the field of view in these videos matched and thus corroborated more precise wind speed measurements taken by the probe. Temperature and humidity sensors also recorded data for about 500 seconds, until they both died in the violence, mud, and debris of the storm.

Tim has not been possessive or proprietary about the design of his probes, despite their success. In fact, when other researchers and chasers have asked him about how he designed and embedded the instrument package, or machined the hardened shell, or anything else about them, he has simply shared the information.

"I'm not one of these guys who says, 'That's my design, and you can't have it.' Look, more data is good. I freely give them all the details. I just ask them to share their data, if they get any," he says. "And I also tell them to be careful."

Tornadoes are so fiercely resistant to human inquiry, and so forbiddingly dangerous, that "tornado networking" among researchers is the best and quickest way to assemble the pieces of the puzzle, for the benefit of science and humanity.

Meanwhile, Tim has started dreaming up another "wild idea" to penetrate tornadoes. He calls it a debris sonde (the word *sonde* meaning "a probe to take atmospheric measurements").

"Actually, it would just be an instrumented rubber Nerf ball tied to a string that snaps when the winds reach 100 miles an hour. Then it would get lofted up to higher elevations and take measurements from inside the wind fields."

If that sounds vaguely familiar (and perhaps a bit humorous) that may be because Tim's Nerf sondes are not unlike the Ping-Pong ball–like sensors dispersed into the (completely computer-generated) tornadoes in the movie *Twister.* One problem with this idea: In order to clarify precisely where in the storm the data were coming from, each

Nerf sonde would have to be embedded with a GPS sensor. But that's beyond the current state of GPS technology; when the balls start to tumble, the GPS sensors get confused about where they are, like children lost in the woods. Nevertheless, Tim believes that within a couple of years, GPS technology will have advanced enough to make this idea work.

Which is part of the reason why, when it comes to designing and deploying tornado probes, Tim Samaras has been the most successful pioneer in the world. He's ingenious, he's bold, he's persistent, he's highly skilled, and he has an imagination that outruns technology. That's about as far from the bored, frightened bureaucratic mind-set that shot down Johannes Peter Letzmann as you can get.

5: TORNADO ON THE GROUND!

"THIS IS IT," MUTTERS TONY LAUBACH DARKLY. "THIS IS THE DAY SOMEBODY GETS KILLED." It's 6:23 p.m. on Friday, May 23, 2008, and the Storm Prediction Center at the National Weather Service has just reported that yet another extremely dangerous supercell, which could spawn yet another extremely dangerous tornado, is bearing down on our location in the tiny town of Quinter, Kansas. It's due to arrive in about 16 minutes.

Tony wrenches our storm-chasing vehicle back toward Interstate 70, intent on blasting a little way west out of town so that we can observe whether this violently explosive storm system will give birth to a tornado without our getting killed by it.

In the chaos and confusion of what has happened in the past ten minutes, Tony and I have become separated from the other five chase vehicles in our group, including the geekmobile, the monster black GMC truck driven by team leader Tim Samaras, which is known on two-way radio chatter simply as Probe and around which the entire TWISTEX expedition revolves. TWISTEX, an acronym for Tactical Weather Instrumented Sampling in/near Tornadoes Experiment, was the name of Tim's overall scientific undertaking, as well as the website *(twistex.org)* where he provided daily updates for everyone actually participating in the field or following along online.

"We are now officially running for our lives," Tony remarks with a kind of unnerving tranquillity as we pull out onto the mostly empty interstate. Around us the great prairie sky is glowering with menace, an immense black slab hanging down out of the heavens. In Quinter all the electricity is out, the rain-slicked streets are shiny, and silvery cottonwood boughs are lashing back and forth in the wind, like wild horses. The eerie wail of a tornado siren echoes down the deserted streets. Behind an immense grain elevator, by far the largest structure in town, the low-hanging cloud bank eerily illuminates from the greenish heights, then illuminates again, like a horror movie. It is as American a scene as anything ever painted by Edward Hopper. It is also as dangerous an American scene as anything out of a high-crime city block. But this is exactly what we have come so far to see and, if possible, to understand.

It was only a short while ago that the Storm Prediction Center issued a PDS bulletin, meaning that what is boiling up into the atmosphere above this particular patch of Kansas is likely to develop, alarmingly soon, into a "particularly dangerous situation."

URGENT—IMMEDIATE BROADCAST REQUESTED—THE NWS STORM PREDICTION CENTER HAS ISSUED A TORNADO WATCH FOR MUCH OF WEST CENTRAL KANSAS AND SOUTH CENTRAL NEBRASKA . . . FROM MCCOOK NEB TO 20 MILES SOUTH OF MEDICINE LODGE KS . . . EMBEDDED SUPERCELLS ALONG WESTERN KANSAS SQUALL LINE . . . FORMING ALONG MERGING COLD FRONT/DRY-LINE . . . DESTRUCTIVE TORNADOES . . . LARGE HAIL TO THREE INCHES IN DIAMETER . . . THUNDERSTORM WIND GUSTS TO 80 MPH AND DANGEROUS LIGHTNING . . . THIS IS A PARTICULARLY DANGEROUS SITUATION.

Despite this sinister warning, the chasers in our group are not particularly daunted by the PDS alert. In fact, the boundlessly cheerful

Verne Carlson, an unofficial addition to the main chase group, who is chasing with his two sons in an ancient Subaru Outback with a quarter million miles on it, cracks over the two-way radio that a PDS is usually "the kiss of death." Meaning that when the SPC claims a tornado is imminent, the storm clouds usually turn to kittens. (For people in the tornado's path, of course, it's a correctly forecast PDS that can be, literally, the kiss of death.)

This time, though, the SPC is right.

Not long after the warning is issued, Tony Laubach's mobile mesonet vehicle, in which I'm riding, tears off toward the south edge of town, where the SPC radar data seem to indicate a tornado is forming. The two other mobile mesonet vehicles follow us. There are now three mobile mesonets on the chase, tagged on radio chatter as M1, M2, and M3—meteorologists Bruce Lee and Cathy Finley are riding in M1; meteorology students Jayson Prentice and Chris Karstens in M2; and Tony and I in M3. The grandly titled mobile mesonets are actually tiny, inexpensive rental cars, dark blue Chevy Cobalts, with mobile weather stations mounted on their roofs. (The pursuit of science always involves the pursuit of money, Tim says, and in this case the team has rented the cheapest vehicles available because money is tight.)

The instrument package looks like some kind of over-the-top high school science project, a sort of doghouse-shaped contraption made of white PVC pipe with a tiny propeller (actually a wind speed indicator, or anemometer) mounted on top.

Now, suddenly, M1, M2, and M3 suddenly skid to a halt at the side of a wet gravel road just south of Quinter.

"Tornado on the ground!" Tony shouts over the radio, vaulting out of the car with a digital camera in one hand and a digital camcorder in the other, multitasking in triplicate.

Now, down over the crest of a low golden hill, perhaps three-quarters of a mile away, is an immense, slate gray tornado, squat and fat and spooky looking. It's not exactly a wedge; it is shaped something like the cooling tower of a nuclear power plant, with a sloping top narrowing to a kind of waist, then widening slightly as it descends. Where it makes contact with the ground, it stirs up huge, boiling veils of black prairie dirt and debris—a particularly unnerving special effect. Off to the right side of the main vortex, to the north, is a narrow, spindly "satellite" tornado, or subvortex, which spins erratically closer and closer to the fat cone, like a wobbly gyroscope, until the two vortices merge.

Storm chasers like to say that if a tornado seems motionless, stirring neither to the left nor to the right, but seems to be getting bigger as you watch it—well, friend, it is heading straight for you. You'd best have an exit plan, and if not, a will. But this wicked-looking tornado is moving slowly to the right, or north, bearing down on the tornado-warned town of Quinter, and the interstate.

The drivers jump back into their mobile mesonets and book north, as the tornado follows along to the west, paralleling the highway. We're keeping abreast of the twister, like cowboys trying to rope a steer, sometimes gaining a little ground on it, sometimes dropping back. Tony wrenches M3 off the road once to snap a few pictures of the thing as it moves across a field, wrapped in clouds of dust. It is now so black it looks satanic. Yet there is something about this furiously spinning whirlwind that is so mortifyingly sublime, something that so transcends this particular day, and this particular place—something that is so utterly *other*—that any comment at all seems ridiculous. Nevertheless:

"Holy smokes, man!"

"Jeez, look at that thing!"

Seasoned observers would later estimate this tornado to be several hundred yards wide and perhaps half a mile away, but to me, a newbie chaser, the thing looks absolutely immense and about 200 yards away. Such are the perceptual distortions of fear and inexperience.

Two great things about storm chasing in Kansas: Because it's so flat and wide open, you can see the entire arc of the sky, which displays summer storms and tornadoes from great distances, as if they were large format films. Second, Kansas is also conveniently divided into mile-square grids, with straight north-south and east-west roads crosshatching nearly the entire state, so that chasers can follow a tornado almost anywhere it goes. (In places where the road network is irregular, or the terrain is difficult, trying to chase a tornado is almost guaranteed to disappoint.)

One problem: Many of these rural roads turn unexpectedly from pavement into gravel or dirt, and during a high-speed chase one may suddenly rocket from smooth tarmac to muddy gravel that's been turned to the consistency of pancake batter by a welter of rain and hail. Crossing from one to the other at high velocity is a good test of hand-eye coordination, the aerodynamics of cheap rental cars, and the quality of a person's theological connections.

In this case, the road turns from muddy gravel back into pavement as we enter the town of Quinter, with the unearthly wail of the tornado siren all around us. Quite rapidly, off to the west, the tornado funnel seems to disintegrate into gray, scudding clouds, a disorganized rabble of gauzy fragments. Then it practically disappears, except for the low-hanging and ominous-looking wall cloud (a low ceiling that sometimes remains once a tornado's funnel disintegrates), which continues moving north.

"Be careful of that thing—it's still live," Bruce Lee warns over the radio. "There's still rotation, even though you can't see it."

Bruce, riding in M1, has a Ph.D. in atmospheric sciences and is overseeing one of the serious scientific aspects of this Kansas joyride. The effort is to drive the three mobile mesonets into extreme close

GLOSSARY

Velocity The quantity that describes how fast and in what direction a point is moving; the speed can be determined by the time it takes a point to move along a certain path.

Barometric pressure The force per unit area exerted by an atmospheric column

GPS (global positioning system) A space-based radio navigation system that allows people connected to the system to calculate their physical position

proximity to a tornado, space them out, and then log data about the storm's wind speed and direction, barometric pressure, temperature, and humidity. Since each vehicle is also equipped with a GPS tracker (which looks like a hockey puck mounted on the dashboard), it will be possible later to pinpoint the precise location of each data station and thus begin to create a picture of the structure of the storm. Tim's efforts to deploy a ground probe are not directly linked to Bruce and Cathy's work, but Tim is by nature a team builder, a collaborator, and every bit of information, from every source, helps to fill in the picture of what is happening inside the tornado.

Bruce Lee and Cathy Finley, as degreed meteorologists, are particularly interested in trying to understand the extremely violent downdrafts (known as rear flank downdrafts, or RFDs) that accompany tornadoes. These winds are not like the benign, high-spirited blasts that can raise an explosion of leaves on a March afternoon; RFDs are so wickedly powerful that they can explode a windshield or knock over a utility pole. They are one of the many ways in which a tornado can kill you.

Right now, though, Tony is more concerned about getting unlost than understanding an RFD.

"M1 to M3, just keep heading north on Castle Rock Road," Bruce is saying to Tony over the radio. "M2, try to space out about half a mile apart."

"M3 to M1," Tony radioes back, "Can't find Castle Rock Road. The road we're on just dead-ends at the railroad tracks on the north end of town, by the grain elevator. Where's the friggin' road?"

"Tony, just keep trying," Bruce continues. "We're heading north up this road. Looks like the core is directly ahead of us, moving north-northeast."

All three mobile mesonets are equipped with wireless Internet access (that is, wireless Internet access when the system has not inexplicably crashed or had a seizure). On the laptop screen mounted between the two front seats, Tony is now furiously multitasking between two programs. One, called GRLevel3, gives continuously updated radar information about the storm cells surrounding us. The other, made by the mapmaker DeLorme, is a program called Street Atlas USA, which shows the grid of streets and highways across the entire United States. He has now zoomed down to one small square of rural west-central Kansas, our current location. It's an amazing, wonderful program, but like everything else created by humans, it has its limitations. For one thing, it does not distinguish between paved and unpaved roads, or indicate where one abruptly switches to the other. And though it is now tracking our vehicle's path via GPS as a fat green line on the map, at this moment we are still slightly confused about our location. We're momentarily lost in the tangle of streets on the north edge of town, transected by the railroad tracks. Finally we locate Castle Rock Road, which drills directly north.

We turn onto the muddy gravel road, and suddenly we are heading directly toward something that looks entirely too much like the unfortunate end of this story. The entire sky seems to have fallen to earth, and lying across the road ahead is a maelstrom of black, boiling clouds and debris sweeping from left to right at alarming speed. Directly in front of the car, tumbleweeds flashing across the road act as handy velocity indicators, and they are now flickering past us at a rapidly accelerating rate.

Ahead of us but out of sight, in M1, Bruce and Cathy are edging up the road, closer and closer to the periphery of the storm. Cathy grew up on a farm in Minnesota, and as a child she was always terrified of big storms. But "fear turned to fascination," she says, and now her fascination is causing her to take chances, drawing her closer and closer to the edge.

The malignant thing that lies across the road less than a mile to the north is so huge and so formless that its outer edges are not visible to us. Several seasoned observers later estimate that it is at least a mile wide, perhaps two miles. It is not, strictly speaking, a tornado—or at least, not *only* a tornado. A tornado is spawned by the magnificent (and rare) atmospheric storm system called a supercell—a sort of thunderstorm on steroids. This massive system, towering up into the stratosphere and filled with explosive energy, develops an area of rotation called a mesocyclone (or "meso," to storm chasers). And it's the meso that actually gives birth to tornadoes. A meso may be many times larger than the actual funnel that drops down to the ground. What appears to have happened on this lonely road north of Quinter, Kansas, is that the entire meso has descended to earth, surrounding and concealing the vortex (or vortices, as the case may be). That's why it appears so immense and shapeless.

GLOSSARY

Rear flank downdraft (RFD) A rush of air down the back side of a storm that descends with or near a tornado

Mesocyclone Referring to the region of rotation between two and six miles wide, from which a tornado might descend; often located in the rear, right region of a supercell

"This tornado is so big it gives me the chills, practically," Bruce says over the radio. "It's just frighteningly large." (As Bruce well knows, this is a meso, not a tornado—but under the circumstances, nobody cares what you call it. Either one could kill you.)

Moments later Bruce and Cathy are given a riveting demonstration of the frightening power of an RFD blast when the utility poles on the left side of the road, already vibrating like mad in the cyclonic wind, suddenly pitch over and fall into the road, with electric-blue power flashes all the way down the line. Bruce wrenches his car down into the muddy ditch in an effort to keep from getting electrocuted. A few hundred yards behind them, Chris and Jayson are unable to elude the falling power lines, which collapse on top of M2, ripping the anemometer off the weather station on the roof of the car. Chris, at the wheel, edges the vehicle out from under the power lines, which bow and pop as they become disentangled from the broken weather station and break free.

Bruce and Cathy are now stuck in a muddy ditch in a tinny rental car, unable to drive backward or forward, trapped between sparking, collapsed utility lines and an oncoming, milewide mesocyclone. This is not the sort of situation that reminds you of slipping off to sleep at home, in your own bed, after a nice hot bath. Cathy later says that in her years of storm chasing this is one of the most frightening situations she has ever encountered. In fact, this is exactly the sort of situation that Tim Samaras, our team leader, has dreaded and sought to avoid for years.

"I really worry about the whole team out there," Tim says. "There are so many dangerous things we could run into, besides tornadoes—and downed power lines are one of the biggies. In this case, everybody did the smartest thing they could do in what was a pretty perilous situation."

Unfortunately, at this moment Tim and the Probe vehicle, along with severe-storm tracker Carl Young and Tim's son Paul, are out of radio contact with the three mesonets.

Bruce, Cathy, Chris, and Jayson (whose vehicle is off the power lines but still parked dangerously close to them) watch helplessly as the boiling meso moves directly across the road, lateral to their position; for whatever reason, it does not shift direction and head straight for them.

Moments later, Tony and I come booming over the hill in M3, heading north on Castle Rock Road. Ahead of the car, perhaps half a mile away, we can see the huge, ugly meso crossing the road ahead, bearing to the northeast. Much closer, perhaps a hundred yards away, we see a small satellite tornado, translucent as a ghost, skipping across the road—a kind of eddy in the cyclonic winds of the main meso. We come upon M2 stopped in the road and M1 down in the ditch. M2 has pulled off the power lines, but its rooftop weather station is badly damaged (though still partially functional). Bruce and Cathy's vehicle is deeply mired in mud off to the right side of the road.

Most important, everybody is safe. And the data loggers in the vehicles (which record information from the weather instruments in the form of great reams of numbers) have captured valuable data about the RFD and the peripheral circulation of a major mesocyclone.

All of these hard-won measurements will be added to what is, so far, a remarkably sparse bank of data about the "tornadic flow field" near the surface of the ground, which, Chris Karstens wrote in a 2008 scientific paper about the Quinter tornado, remains a "mysterious and harsh environment." Measurements of the RFD blasts and other ferocious winds in and around the tornado could be of particular value to structural engineers seeking to create buildings that would withstand the phenomenal stresses of tornadoes, Karstens wrote. Even the RFDs can be amazingly destructive: In another paper about the Quinter tornado, also published in 2008, Bruce and Cathy reported that M2 had recorded a peak RFD wind speed of 103 mph, equivalent to an EF1 tornado, and M2 recorded gusts even higher just before the power lines blew down on it.

In the distance, out across the prairie grasslands where such remarkable things descend out of the summer sky, the horizon has already begun to brighten. As the meso moves off to the northeast, daylight appears along the rim of the world, Kansas's wraparound view of heaven. There is the sound of the ceaseless, sibilant rush of a light breeze stirring the wet grass.

The other sound one hears is the excited chatter of storm chasers, who now begin to gather around the wrecked cars. A young guy comes running up with a big digital camcorder. He tells Jayson that he got the power pole collapse on video. Chris Collura, a Florida-based chaser, pulls up in a rental car with Illinois plates, which has one side window completely blasted out by the RFD. He had been videotaping from farther back down the road, and because of his broader perspective, he captured something none of our group had seen: Before the immense meso touched down and enveloped everything in sight, he'd seen an enormous wedge tornado considerably to the west. This brings home another of the manifold dangers of storm chasing: A tornado can be rain wrapped, meso wrapped, or hidden in any number of ways. (In this case, the tornado was hidden both because it was well to the west and because our view was blocked by the meso.) You can't tell exactly where it is until it's right on top of you.

Just then Tim pulls up in the mud-slathered Probe, the mother ship for this whole expedition.

In addition to Tim's predilection for engineering, electronics, and meteorology, other, more down-to-earth skills become evident when he jumps out of the geekmobile near where Bruce and Cathy have run off the road. Tim's truck has a heavy-duty winch on the front end, so Tim drives close to the mud hole where M1 is stranded, hooks the winch to the little car's frame, and yanks it out of the ditch. Unfortunately, the winch also yanks a spring out of M1's suspension, rendering the little Chevy Cobalt about as useful as a dead duck.

"Man, we got a great view of this amazing transformation from a 'stovepipe' [a pipelike tornado, straight up and down on the sides] into that huge wedge," Tim tells Bruce as he's standing there in the muddy road, trying to figure what to do next. "Looked like the whole meso just dropped right down to the ground."

Meanwhile, Carl Young, the crack meteorologist, has been analyzing the latest radar data from the SPC, which is streaming in over Threat Net. The news looks great—or terrible, depending upon your point of view.

"Looks like there's another supercell with a pretty massive hook in it moving towards Quinter," Carl tells Tim uneasily. "Looks like it should be here within 30 minutes."

On radar, tornadoes often produce what is known as a hook echo, because the region of severe weather and heavy rain at the core of the storm has the appearance of a fat fishhook on the computer screen. This news brings out another of Tim Samaras's special skills. He greets

Carl's warning that our current location is squarely in the crosshairs of an onrushing tornado with Buddha-like serenity. (Actually, Tim confides later, he assumed Carl was just saying 30 minutes to soft-pedal the danger of the situation. Tim figured the tornado would probably be arriving a lot sooner.)

Tim is like the still point at the center of the storm, projecting composure and competence to the swirling vortex of people, projects, and equipment around him—in this case, 15 people in six different vehicles, including the three mesonets, a National Geographic TV film crew, and Verne Carlson and his two tornado-chasing sons. Taken together, this group constitutes the ever shifting TWISTEX team.

Standing in the muddy road beside the geekmobile and the newly rescued, but disabled, M1, Tim is now focused on the new, annoying, and slightly absurd problem that has just been created: How do you compress a heavy-duty car spring and then squeeze it back in place in order to fix a car in order to flee a tornado? It's clear that he is accustomed to just this sort of situation—repairing weird problems out in the field, with too little time, without the right tools, moments before the hammer falls. In fact, he says later, he's been doing this kind of stuff for the past 30 years, except that usually there's a client with a multimillion-dollar investment in the outcome.

By contrast, this is easy. All you've got to worry about are 200-mile-an-hour winds and a little mud.

He yanks the car spring out from underneath the disabled Chevy and then gets somebody to stand on it, attempting the quick and dirty solution by trying to lash the compressed spring tight with a short length of clothesline. No go. Then he tries doing the same thing with two extension cords. No go. *Damn.*

Despite the danger, discomfort, and grittiness of the situation, Tim seems right in his element, squatting in the mud in a ditch in Kansas, looking up underneath the stranded Chevy, while the National Geographic film crew get it all down for a National

Geographic Channel show about him airing in 2009 called, appropriately, *Master of Disaster.* Finally Tim hits on the idea of using the weight of the car to compress the spring, then lashing the squished spring tight with the two extension cords. He does that, then he takes the compressed spring and wedges it back into the Chevy's suspension. Then he cuts loose the extension cords with a pocket knife. It's crude and it's ugly, but it works. OK: Good to go.

"Tim's a genius," Tony says later. "He's like MacGyver in that old TV show—the guy who could defuse a nuclear bomb with a Swiss Army knife and a peashooter."

But now the new supercell is bearing down on us, right on schedule. Tim, Carl Young, and Tim's son Paul jump into the geekmobile and take off, hoping to get in front of the tornado and deploy a probe. Tony and I and the other mesonets also head east, following Tim, to the town of Hays. By now the world is growing dark, with occasional glimmerings of an eerie pea green light not unlike the light that probably filters down from Oz. In the little towns along the interstate you can hear tornado sirens wailing like lost children. Behind us, when the lightning flashes, we can see what a hook echo actually looks like when you're close to it. It looks like a big black wall of death. A couple of ambulances flash by us, heading east, sirens wailing. Somebody's in trouble up ahead. It's now 6:23 p.m.

"This is it," Tony says. "This is the day somebody gets killed."

(And, in fact, the next day it is reported that two motorists from Colorado were killed when their vehicle was swept off the road and flipped over along the very stretch of interstate we've been crisscrossing all night.)

We stop at a gas station to hurriedly fuel up. There are three or four carloads of people parked in the covered bays, apparently seeking shelter from the tornado. The protection offered by the tin roofs overhead is absurd, akin to holding a paper plate over your head.

"Says on the radio somebody saw a mile-wide tornado to the east of us," a guy at the gas station tells us.

Radar can show the hook echo of a tornadic storm, and Doppler radar can detect directional rotation, even what is known as a tornado vortex signature, or TVS. But all of this amazing electronic information is only circumstantial evidence—it is not *proof* that the storm cell has actually produced a tornado. The only certain proof comes from a direct sighting by a reliable eyewitness.

So now we hop back on the interstate and head farther east, all the while monitoring radar reports on Mobile Threat Net and GRLevel3. Ahead of us the sky is so dark it seems to be absorbing light like a black hole. At the horizon line, there is a faint, eerie, distant glimmering. The interstate is slick and shiny with rain. Downbursts of rain and occasionally hail hammer the windshield. The TWISTEX team has become disorganized and separated, and with the two-way radios operational only at about a mile's distance, we've lost touch with Tim again, as well as Verne and several others. There are cars stopped, warning lights blinking, along the shoulder in both directions. The mile-wide tornado described by spotters may very well just have passed across the Interstate.

Tony radios Tim, whom we have not heard from in too long.

"Probe, this is M3. Are you there?"

Silence.

"Probe, this is M3. Tim, where you at, man?"

No answer.

"I just hope whatever just went through here didn't get him," Tony mutters.

Later we learn Tim, Carl, and Paul's side of the story. They stopped in Quinter to deploy a couple of probes before the second powerful tornado crossed the interstate a couple of miles west of town (missing the probes completely). Still, they came within about 100 yards of a massive wedge tornado as it crossed the road ahead of them. In the darkening gloom, shot through with lightning, they could see the tornado moving north and then beginning to dissipate; then they could see a "tornado handoff" taking place—a new tornado being formed as the old one died, a whirlwind born of a whirlwind.

"The roads were terrible, and it was getting dark fast, so we gave up the chase," Tim said later. "Then, on Threat Net, we could see this massive hook echo moving up from the south—an enormous supercell which probably had a tornado embedded in it. It was heading straight for the town of Ellis, about 40 miles east on the interstate.

"We hauled butt down I-70 to Ellis, where we stopped and waited for the storm to come through. On radar the core of this thing was just absolutely massive. As the hook echo approached, we tried to sample hail, but it was just little stuff, nickel to quarter size. Then, as this intense hook echo approached, we could see power flashes from falling power lines to the west of town. We could see another massive tornado just to our west, or at least kind of see it, every time there was lightning or a power flash. Then—*boom!*—we started getting hit by RFD blasts as the tornado approached, so we drove a few miles farther east, then stopped again to watch yet another tornado moving through town. We saw another string of power flashes, then the entire town of Ellis was plunged into darkness. The tornado seemed to dissipate as it approached I-70. I wish I'd deployed a few probes in Ellis, but there were a few small inconveniences in the way of that plan—like RFDs, the fact that it was pretty much dark downed power lines, and another tornado."

Tony and I and the other mesonets also head east on the interstate, attempting to find a location that is close but not too close to the periphery of this horrendous storm system. On the radio, Carl reports that the SPC is reporting CAPE values of over 4,000 joules per kilogram in this area. That's fantastic. CAPE stands for convective available potential energy. It's a complex equation measuring temperature, moisture, and elevation, but what it means in plain English is how much rocket fuel is available to stoke a tornado. A CAPE value of

1,500 to 2,000 means there's enough fuel to power up a decent twister. But a CAPE value of 4,000 is like topping off a drag racer's tank with high-octane boogie juice. A CAPE value of 6,000 is as high as the SPC charts go. Anything above that is literally "off the charts."

"It looks like this system is gonna keep on truckin'," Carl reports, exultantly.

At 7:50 the SPC reports a "large and extremely dangerous tornado just south of Cedar Bluff," ten to fifteen miles south of where we are. The radar also shows a menacing storm cell moving toward Greensburg, Kansas, much farther to the south of us. Greensburg was the town that was virtually wiped off the map the previous summer, on May 5, 2007, when an enormous EF5 tornado hit the town after dark, leveling practically everything except the grain elevator, city hall, and a liquor store. Nine people were killed.

"The damage in Greensburg was about as bad as it gets," Tim says. "The town was basically flattened—it's amazing only nine people got killed. But maybe someday our work will provide five or ten minutes' more warning time for people. That would be huge."

By now, it is getting dark. A certain weariness has begun to set in, a weariness of sustained adrenaline rushes, high anxiety, long days, and bad road food. The storm system has also gotten so sprawling, and so confusing, it's difficult to follow what's going on. By the end of the evening, the film crew's van will have had two windows blown out after a close encounter with an RFD, and two tornadoes will have been sighted in Quinter, two in Ellis, one in Cedar Bluff, and one very small one down in Greensburg.

Referring to tornadoes, one weary Greensburg resident tells the local paper the next day: "I just wish they'd go away and stay away. "

It's after 10 p.m. when we all converge in Hays, where we've rented a block of inexpensive hotel rooms. The group of us go out for a bite to eat and discover practically the only place open after the storm is an

Applebees restaurant. The place is packed with storm chasers, most of whom know each other. (Tim estimates that the entire community of hard-core, more or less full-time storm chasers amounts to only about a hundred people; the semiregulars, who love to chase when they can, or regularly go on commercial chase tours, number only about 300 or 400.)

Spirits are running high on this amazing day—the TWISTEX team has seen seven tornadoes and gotten data on three. Chasers start yanking out their digital cameras, digital camcorders, and laptops, showing off the tornado images and footage they got today (sometimes called torn porn). It's like a 21st-century version of gunslingers in the Old West, each chaser trying to outgun the next one. Tony whips out his Nikon and shows off his images of the black, satanic cone (the one that looked like a nuclear power cooling tower) we saw south of Quinter earlier in the day. Killer. Another guy shows off a glowering image of a wedge tornado wrapped in black veils like the angel of death, just scary as hell. Gnarly.

Then a big swaggering guy wearing a ball cap and a T-shirt, probably six two, 240, saunters up to the bar, slaps down his laptop, and lets 'er rip. The guy's videotape, clearly shot with a dash cam (a camcorder mounted on a vehicle's dashboard), shows a scene that we immediately recognize as Castle Rock Road, north of Quinter. There's no sound on the audio track except the *slap-slap* of the windshield wipers and the lashing of rain. The gravel road ahead of his vehicle is completely empty, with a roadside windbreak of cedar trees whipping furiously in the wind and the monstrous meso boiling up into the sky to the left side of the road. There are no other vehicles visible. So he's the guy who got the closest, considerably closer than Bruce, Cathy, Chris, and Jayson. In fact, as the video runs, it's apparent that he is almost insanely close. The guy is out there alone, about to be consumed by a milewide mesocyclone, but he's not saying a word. Once or twice he coughs. Then you can see a huge wedge-shaped funnel, low and squat, forming very close to the left side of the road. The cone rapidly shape-shifts into an elongated, eerily beautiful, whitish funnel with great skirts of dust swirling around its base. Then it begins to bear down onto the road. It couldn't be more than a hundred yards away.

At this point, the driver jams the vehicle into reverse and starts backing up the road, with the dash cam still running, as an immense rotating mass of grayish brown dust, vapor, and debris engulfs the funnel and then begins swallowing up the place he has just vacated only seconds before. Finally the driver speaks:

"We're about to get hit by a friggin' wedge!"

You can see straw-colored tumbleweeds flashing by in the headlights, left to right, so fast they register as no more than blips in the brain. The mesocyclone has turned into an enormous, shapeless, roiling mass, filling everything in sight. The driver speaks again:

"Wedge tornado, a hunnert yards in front of us!"

Then power lines collapse into the road ahead, and everything begins to go dark. *Slap-slap-slap. Cough.* Then it's over.

When the guy's video ends, some chasers at the bar actually applaud. If this is a shoot-out, the Dash Cam Kid won. The big guy mentions that he sold his footage to the Weather Channel this afternoon, and they've been airing it nationwide all day. Then he saunters away.

"Doug Kiesling," Tony says. "On the Web he calls himself Lightningboy or the Weather Paparazzi. He takes crazy risks, but he definitely gets the shots."

In the chase community, Tony says, Kiesling's reputation is much like the performance he just gave at the bar: showy, self-promoting, and competitive. He is not interested in science, only in getting the best footage. The tagline of his website *(www.lightningboy.net)* is: "I risk my life for the shot so you don't have to." But despite all that, Tony adds, when he worked as a stringer for the severe-weather video service Kiesling runs, called Breaking News Video Network, Lightningboy actually turned out to be hardworking, fair, and honest. Even so, the risks he took today to get that videotape are so close to the edge, so perilously poised between the narrow margins of error, one has to wonder if it's worth it.

Tim Samaras and his TWISTEX team have an entirely different agenda. He is fascinated by the goal of penetrating the mystery of tornadoes, but only in order to understand them for the benefit of science and the safety of humanity. He takes calculated risks, but only on behalf of human knowledge and safety. He has no interest at all in needlessly tempting the fates, or outgunslinging anybody.

"I am not out here to take foolish chances," he says. "I am not a wild man. If the road is not passable—if it's too slick or too muddy or it just looks too dangerous—I won't go in. I take total responsibility for all my crew members. I feel very responsible for every member of my team."

Even so, there is simply no getting around the fact that tornadoes are the most violent storms on Earth, and there is just no way to learn about them without taking the kinds of risks some reasonable people might consider suicidal. Mathematical modeling on supercomputers will get you only so far. You have to meet the beast face to face. Tornadic winds have the most extreme velocities ever measured on the planet—in one documented case, a tornado near Moorhead, Minnesota, actually picked up a train car containing 117 passengers, with a total weight of 83 tons, and tossed it 80 feet. If you want to understand how such a thing is possible, you need to get uncomfortably close to the thing as it happens. If you don't want to understand, good: Get a job at a bank.

Tim Samaras is entirely aware of the dangers inherent in what all these people in this noisy Kansas bar are doing, including members of his own team.

And it worries him.

"Someday," he says, "somebody's gonna get bit."

6: A MILE-WIDE, SOD-SUCKING MONSTER

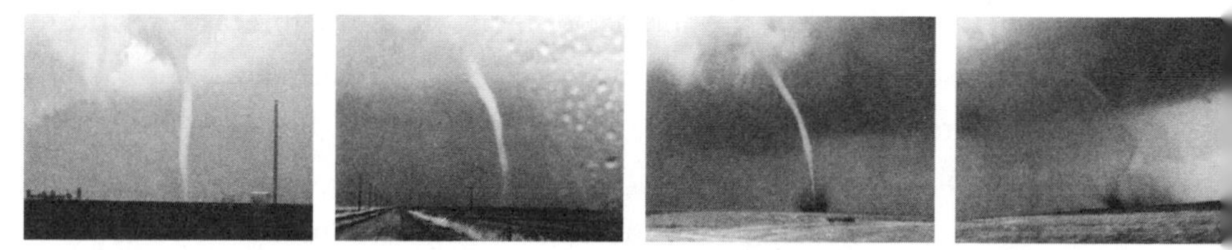

CHASE DAYS DO NOT ALWAYS START OUT WITH A STORM PREDICTION CENTER "DAY One" forecast that's full of the promise of chaos, madness, and mayhem. Sometimes they start out with a prediction of bland blue skies and sunshine—the sort of bitter disappointment that storm chasers experience entirely too often. Thursday, May 22, 2008, was a day that started out that way. But it certainly wasn't a day that ended that way.

The previous day, a Wednesday, Tim had sent out his daily TWISTEX e-mail alerting the team that something promising was beginning to take shape in the atmosphere over northwest Kansas. He wanted everyone to converge in Denver, where he lives, to get ready for action. His e-mail, as usual, was calm and understated:

> **RUC is indicating very poor moisture return at 0z, thus there will be no TWISTEX operations today. Tomorrow, moisture improves significantly, along with a very favorable upper wind profile, thus we'll be GO for operations tomorrow—likely along the triple point somewhere in NW Kansas. Future holds good promise the next several days through the holiday weekend. Please be prepared for a multi-day mission. . . .**

The RUC (pronounced ruck) is the National Weather Service short-term forecast, issued every three to twelve hours. The acronym stands for Rapid Update Cycle, and because of its short-term nature it tends to be a bit more accurate than other forecasts. Nevertheless, chasers know enough to sometimes make fun of the RUC because, like every other forecast, it can also be wrong or misleading.

The presence or the absence of moisture in the forecast is critical, because it's moisture that fuels tornadoes. If tornadoes are a kind of spiraling conflagration, then moisture is the fuel that keeps the fire going.

The term "0z" (pronounced zero-zee) does indeed look like the name of the place to which we do not wish to be blown, along with Dorothy and Toto. Actually, 0z refers to the time as denoted in Coordinated Universal Time, abbreviated as UTC but often called, simply, Zulu time. Scientists tend to use UTC because it's the same all over the world and therefore absolutely clear and unambiguous. It does not involve translating into eastern standard time, central standard time, or any other kind of local time. In Colorado, 0z corresponds to 6 p.m. So the short-term forecast (the RUC), as of 0z, means the Storm Prediction Center's forecast as of 6 p.m., Tuesday, May 20, 2008.

The triple point refers to the fact that three atmospheric phenomena are converging in the same place, producing an area of extreme atmospheric turbulence, ripe for tornadoes. One is a warm front, one is a cold front, and the other is the dryline. The dryline refers to a massive, semipermanent feature of the Great Plains, and the reason why this area is the world's hottest hot spot for tornadoes. From the south, great tsunamis of warm, moist air move north out of the Gulf of Mexico. From the west, huge waves of cool, dry air move over the Rocky Mountains, heading east. Where these two air masses converge (the dryline), the atmosphere tends to create massive instability which is very likely to produce tornadoes.

So far this season has been extraordinarily active—in fact, 2008 is on track to become the most active tornado season on record. Before the week is out, there will have been 110 fatalities reported nationwide after massive, widespread tornado outbreaks. Normally May and June are the TWISTEX team's chase months, and we're not even to the end of the first month.

In the Denver hotel room on Wednesday morning, waiting for Tim and his group to show up, Tim's meteorologist sidekick Carl Young pores eagerly over the incoming data on his laptop. Carl seems to have an almost unlimited capacity for poring over data, as if they were something yummy to eat or drink. He's very excited about our prospects today, especially the area around Great Bend, Kansas. That's about 300 miles due east of Denver—nothing to Carl and Tim, who will drive an average of 30,000 miles a chase season and change the oil in their vehicles every three or four days. With gasoline prices teetering on the brink of four dollars per gallon, and the geekmobile getting all of ten miles per gallon, just gassing up this expedition is like taking out a mortgage on a small home. (This summer the TWISTEX team gas bill will top $20,000.)

GLOSSARY

Triple point The intersection point of at least two boundaries, and often the focal point for thunderstorm development

Dryline A boundary that separates warm, dry air from cool, moist air and is usually where thunderstorms form

Convective available potential energy (CAPE) The potential energy of a storm system, measured in joules per kilogram, used by storm chasers and weather experts to predict how the storm will react; more CAPE means that the storms could be more severe.

Carl says the CAPE (convective available potential energy) values—a technical term that essentially means "tornado fuel"—look great over west-central Kansas. That means there is an enormous amount of available warm, moist air in the atmosphere. He is also studying another piece of juicy data, called the hodographs of the area. The hodographs show, in chart form, what the wind is doing as it gains elevation. Directional wind shear is one key ingredient in the creation of a tornado, meaning that, with increase in elevation, the direction of the wind tends to spiral around. At Earth's surface, the wind might be blowing from the east; at 2,000 feet, from the northeast; at 4,000 feet, from the north, et cetera. This spiraling increases the likelihood that an ordinary, static supercell may tip over into rotation. What's even better for tornado formation than directional wind shear is speed shear—meaning that the velocity of these spiraling winds is likely to accelerate as they gain elevation. This tends to create a rolling effect, like a log rolling in a stream; it is believed that this rolling parcel of air, which can be horizontal to the ground, is then tilted into the vertical, where it becomes the fearsome mesocyclone at the core of a tornado. If all or most of these elements are in place, they'll show up on the hodograph as a swooping fishhook shape. Now Carl, studying the hodographs on his laptop, has a kind of weird gleam in his eye because he's seeing beautiful, near-perfect hooks all over west-central Kansas.

GLOSSARY

Directional wind shear The element of wind shear that is due to the change in wind direction with height

Speed shear The element of wind shear that describes the change in wind speed with height

Gustnadoes Small, extremely weak tornadoes that often appear as only dust whirls near the ground

"I wouldn't want to be flying over Kansas today, with all that wind shear in the atmosphere," he says.

Outside the hotel, huge cumulonimbus clouds have begun piling up in the sky into storm towers, like florets of cauliflower in some storybook world where people are the size of ants. In fact, the magnitude of the atmosphere, especially out here on the Great Plains, truly does dwarf human life—the storm towers of an angry supercell can blossom up to 50,000 or 60,000 feet, more than twice the height of Mount Everest. Maybe that's partly what attracts storm chasers: In a world in which we generally experience nature in a form that is prettily packaged, manicured, reduced, condensed, neutered, trademarked, and boxed in, out here you can see nature properly—magnificent, boundless, uncontrollable, and terrifying.

A MILE-WIDE, SOD-SUCKING MONSTER

One by one, the various team members have converged in the parking lot of this modest hotel in Denver, not far from Tim's home in Lakewood. People have flown or driven in from considerable distances—Carl from his home on Lake Tahoe, California, meteorologists Bruce Lee and Cathy Finley from Minnesota, meteorology students Chris Karstens and Jayson Prentice from Iowa, I from Virginia. What's shaping up in the atmosphere to the east of us looks so explosive and volatile that everybody is hoping it will be worth the trip—maybe we'll be able to deploy a probe right into the heart of this thing, like Ahab spearing the whale.

Carl will be riding shotgun in the geekmobile with Tim and me; Tim's son Paul with Tony in M3 (though later we'll swap seats); Chris and Jayson in M2; Bruce and Cathy riding together; and Verne Carlson with his two sons taking up the rear in his old battle-ax Subaru laden with model airplanes he hopes to fly into the core, plus the old piece of broomstick handle he uses to prop open the hatchback. Various other chasers will be joining up with us along the way.

We head east toward Kansas. Tim, normally calm and level, is excited.

"Maybe we'll see a milewide, sod-sucking monster," he says, clearly implying by his tone of voice that this would be something dearly to be wished for.

Ahead of the geekmobile, the sky is blue and empty except for a few scudding high-level cirrus clouds and the occasional turkey vulture, wheeling on the updrafts. Meanwhile, behind us, the piling clouds continue to mount in the summer sky, occasionally shot through with a jagged shaft of lightning.

Once or twice Tim glances out the side window, to the north and behind us, when he spots gustnadoes, which are weak, short-lived vortices akin to dust devils, spinning up in the fields. "Maybe they're just 'tractornadoes,'" he says dismissively, using a joking term for dust-ups caused by farm machinery.

By 10:30 a.m. we have crossed eastern Colorado and are nearing the Kansas line. The landscape is flat, dusty, dry, and a bit melancholy. It makes one long for the city. A forlorn, broken road sign advertises "ABLE TV, TERNET." There are tumbleweeds trapped in barbed wire fences. Another sign says "See The Biggest Prairie Dog In The World. Hays, Kansas." It seems a trumped-up promise, a genuinely lame reason to keep going.

Then we pass into Kansas. A newbie chaser, observing that we are heading toward the wide-open blue skies of Kansas while immense storm clouds are piling up behind us, asks why. Tim explains that, in essence, there just isn't enough rocket fuel in the atmosphere around Denver to stoke a decent storm. The CAPE values are pretty weak, the wind shear profile is not favorable. By contrast, the big boys are out in Kansas. The SPC's "Day One" says that there is a high risk of tornadoes in the area where we're headed, plus a 30 percent chance of large hail and damaging winds of 50 knots or higher (or close to 60 mph).

OUTBREAK OF SUPERCELLS AND TORNADOES CONTINUES AHEAD OF DRYLINE . . . HIGHEST PROBABILITIES TOWARD CORRIDOR FROM SOUTHWESTERN NEBRASKA SOUTHWESTWARD ACROSS WEST-CENTRAL KANSAS . . . MOST DANGEROUS AREA BEFORE DARK SHOULD BE WARM FRONT ZONE IN NORTHERN KANSAS . . . POTENTIAL FOR INTENSE/SUSTAINED LOW LEVEL MESOCYCLONES AND SIGNIFICANT TORNADO FORMATION . . . .

To storm chasers, an SPC "Day One" like that is as good as getting a love letter from your sweetie. Maybe better. For local residents in the crosshairs, it's time to make sure you know where your loved ones are, turn on the Weather Channel or local radio, and make sure you know where you're going to hide if it hits.

It's shortly after 11:30 a.m. when Tony Laubach in M3, coming along behind Tim, gets a disconcerting message on his laptop. The

SPC has reported that there is a spotter-confirmed tornado just north of Denver.

"What?" he explodes, scanning the SPC radar map on GRLevel3, which is running on his laptop. *"That's near Gilcrest. GILCREST!"*

A couple of minutes later, at 11:42, he radios to the others, incredulously: "There's a quarter- to a half-mile-wide tornado on the ground in Gilcrest! *That's ten minutes from my apartment!"*

He slams the steering wheel with an open hand.

"Damn! What am I doing here?"

Text messages start pouring in on his cellphone, from friends and colleagues, wondering if he's on this one. Tony maintains a lively storm-chasing website *(www.tornadoeskick.com),* which got 80,000 hits last week. His clunky little chase vehicle is in fact dragging along behind it clouds of virtual followers, like seagulls tailing a shrimp boat—tens of thousands of people who would love to come along, but for whatever reason—time, opportunity, money, or fear—cannot.

"Damn!" Tony says again, tasting the bitter dust of disappointment. *"Damn!"*

He calls a chase partner in Denver who also works as a stringer for Channel 7 and tells him to charge over to Gilcrest to see if he can catch this thing. If it's too late, he can at least do a damage assessment.

"I'd do it myself, but I'm on I-70 in friggin' Kansas, headed east!" he yells into the phone. "I'm too friggin' far away!"

Then he reads aloud from the latest SPC severe weather update:

NWS DOPPLER RADAR IS TRACKING A LARGE AND EXTREMELY DANGEROUS TORNADO NEAR GILCREST COLORADO . . . APPEARS TO BE TARGETING FT COLLINS . . . SIGNIFICANT DAMAGE HAS BEEN REPORTED. . . .

"Damn, damn, damn!" Tony howls. "Man, I sure hope we find some tornadoes in Kansas, or this is gonna seriously bite! We have to score today. Man, this hurts!"

Unfortunately, the radar north of Denver now shows a classic hook echo and a tornado vortex signature (TVS), almost certainly indicating a twister where there was not supposed to be one. Nobody predicted this, not even the SPC. In fact, Tony will point out later in an attempt to console himself, the SPC issues "probabilities" or "probs" of various anticipated weather events, and in this case they did not issue a tornado prob of 5 percent or even 2 percent, the lowest level, for Denver. This was a fluke, an almost completely unanticipated event. It was just one of those things that make you realize how complex the weather is—and how much of a role luck can play in storm chasing.

That's why almost every storm-chasing vehicle, no matter how laden with scientific instrumentation, also contains some small object that's the equivalent of a lucky charm. Tony likes to wear an old Washington Redskins jersey bearing the number 81 (wide receiver Art Monk's number) because he was wearing that jersey the day he ran into an incredible string of twisters. Tim Samaras has an ancient cheeseburger on the dashboard of the geekmobile because he had it there just before he ran smack into a fusillade of tornadoes one day years ago.

Such unscientific talismans may seem silly, but they're a reminder of the extraordinary complexity of atmospheric processes, and of how rapidly they can change. It's no surprise that modern chaos theory grew out of meteorology. There are, in fact, three degreed meteorologists in the TWISTEX team, using the most advanced and up-to-date radar available, and we have completely missed what is turning out to be one of the most significant tornadoes in Denver in the past century. It brings home not only how difficult it is to predict these things but also how important—the average warning time for tornadoes is between ten and thirteen minutes. Even an additional two or three minutes' worth of understanding could save an enormous number of lives.

Still, all this doesn't make the present situation hurt Tony any less. He is practically jumping out of his skin. This is what he lives for, and he's heading in the wrong direction.

"Man, I should take up knitting! Tae kwon do!" he howls, pounding the wheel.

At 12:37, the SPC reports that a tornado has been sighted on the ground near Gilcrest, and that a tornado warning remains in effect until 12:45 in east-central Larimer County.

Tony has trouble remembering his mother's birthday, but he has instant recall of all his most memorable tornado sightings. On May 29, 2004, in Conway Springs, Kansas, he saw five in a row. "It was like a satanic dance," he says. He called his video of that sequence, an EF3, "The Devil's Dance," and had a friend compose techno music to go along with it.

On April 13, 2007, the SPC sent out a "High Risk Day One" on an area near Dallas. Magnificent storm towers started going up early in the day. Other chasers called it a sucker storm and went off after different storm cells. But Tony stuck with the Dallas storm, which turned into a half-mile-wide tornado, "hauling butt . . . it was beautiful." Now, he says disconsolately, if he had just slept in and stayed home in Denver, today might have entered his personal hit parade of most memorable storm-chasing days of all time.

At 1:15 the first fatality is reported. There are reports of cars having been picked up and thrown a half mile into a lake. *USA Today* is now reporting that the tornado tipped over 15 freight cars.

At 1:45, Tony starts watching streaming video from Channel 7, in Denver. There are pictures of houses with the roofs knocked off, roofless bedrooms gaping at the Colorado sky like topless shoe boxes. Forty thousand people are out of power in Windsor, where there has been huge damage. Public schools are now on lockdown, and parents are

GLOSSARY

Hook echo Also called a hook, the characteristic shape of the region of precipitation that follows behind a storm and wraps around the updraft, as seen in radar images

being warned that it is not safe to pick up their kids. The preliminary damage estimates put the tornado at an EF2 or EF3. There are warnings that there may be more fatalities.

Tony is going crazy now:

"I'm going to sell all my gear and become an organic farmer! A bass fisherman! A basket weaver!"

Our convoy of six vehicles stops somewhere in Kansas to fuel up at a truck stop and grab some storm-chaser food selected from the four basic food groups: trail mix, caramel corn, crackers, and soda. Occasionally the odd living thing—an apple, say, or an orange—is added to this diet. When we climb out of the vehicles, people crowd around a laptop to watch the live video feeds from Channel 7, showing tremendous tornado damage in Weld County, north of Denver. Tony just shakes his head and walks briskly around the parking lot, as if he has to pee very badly.

But here, in Kansas, it is immediately evident that the atmospheric situation on the ground has changed. Almost unnoticed, over the past hour the blue Kansas sky has darkened, and now slate gray cloud decks are scudding by overhead. The wind is booming through the utility lines; the utility poles themselves shimmy and vibrate alarmingly. This looks to be inflow to a storm system up ahead, Tim says—a hungry supercell, or supercells, inhaling hot, wet air into its ten-mile-high lungs. Things are starting to look up. Maybe Tony shouldn't stock up on bass-fishing gear quite yet.

As we pull away from the truck stop, an immense tidal wave of rain lashes the cars, then briefly turns to nearly golf ball–size hail. Tim still has his hailstone-collecting contraption mounted on the roof of his truck, but these hailstones are puny, beneath contempt; what he wants to find are the big boys—softball-size or even larger, hailstones that could actually kill you.

Tony grew up in the little town of Circleville, Ohio, and clearly remembers the first tornado he saw—on television—when he was ten years old. It was the Andover, Kansas, tornado of April 26, 1991, and he still has the old VHS tape he made of the TV reports on that day. He was riveted. He couldn't get over it. He started ravenously consuming all the books and tapes he could find about tornadoes, like a man who did not previously realize he was starving to death. Then one day there were tornado warnings not far from his home in Circleville.

"Want to go chase it?" his dad asked. It was like asking a duck if it wanted to swim. He and his father, a medevac helicopter pilot who loved nothing better than a good adrenaline rush, jumped in the car and booked after the tornado. Tony drew a picture of a supercell on a map and they charged after it, punching through the "hail core" that he had read about, and then, to his absolute amazement, coming face to face with the tornado itself. Both he and his father were mesmerized. They chased the dissipating dust whirl of the funnel for ten or fifteen minutes, until it disappeared.

"It made it so special that my dad was there with me that day," Tony remembers. In fact, "when we got home, my dad was bitten by the bug worse than me!"

He lived in Circleville for 13 years, and the year he moved to Denver, an F3 came through and devastated the town. The bitter sting of disappointment Tony felt then was akin to what he is feeling today. It wasn't that he wished bad things on his hometown. It was, in fact, that he loved his hometown, and wanted to understand the black whirlpool that had descended out of the sky and destroyed it.

That's partly why he was now completing a meteorology degree at Metropolitan State College in Denver—and why, by the end of this trip, his ravenous curiosity will have led him to witness more than 100 tornadoes. Maybe someday he'll have seen a thousand. If he could have one of those Make-A-Wish Foundation wishes granted, he says, it would be to have this sign posted at the edge of his old hometown:

Circleville, Ohio. Home of Tony Laubach, Storm Chaser.

It's 2:40 p.m. when the SPC announces that there are tornado warnings down near Scott City, Kansas, on White Woman Creek, about 30 miles south of I-70. That's in front of us, not behind us. The storm cell is moving north at about 45 to 50 miles an hour—"haulin' butt," as somebody says—so it looks as though we will intercept it as we move east on the interstate. Tim steps on the gas, heading toward Colby. Around us the fields are golden and glowing in a kind of archangel light, with tiny, toy black cattle and a black, towering sky. We pass a sign that says "Welcome to Colby, the Oasis on the Plains." We stop briefly at a truck stop to repair one of the mobile mesonets with duct tape, beneath a bizarre oasis formed by preposterous steel palm trees supported by guy wires. In the rising wind, the palm fronds make an eerie sound akin to the creaking of a steel ship at sea. By now, Channel 7 is reporting 100 injured in the tornado that rampaged through Gilcrest, Colorado.

On one of the Internet programs the team is monitoring, it's possible to locate other storm chasers, who show up as tiny icons on the radar. By dragging the cursor over the icon, the chaser's name and other identifications become visible. The program now shows that there are 36 other chasers on the storm cell we are pursuing. We are not alone in having chosen west-central Kansas today, rather than Denver.

"Tornado spotted nine miles north of Monument, Kansas, moving north at 44 miles an hour," Jayson reports over the radio from M2, reading from an SPC alert. That's very close to I-70. It has begun to rain, a hard, gray, driving rain. We get off the interstate at Grinnell, where a sign boasts gaily, "Try the C. W. Parker Carousel, a National Treasure." We are searching for a road south and then east, to intercept the north-moving storm cell, but we're trying to avoid dirt roads due to the rain. Paved roads are difficult to come by this far out in rural Kansas. And overhead, to Tony's alarm, the sky seems to be clearing. The great, packed cloud masses of the storm cell are breaking up into bulbous, inflated, slightly humorous clouds, like children's toy boats floating off in all directions.

But Carl Young doesn't care. He sees on the radar that something fantastically interesting is heading our way.

"We have to wait for the precip to go over to see inside the hook, but man, what a hook!" he enthuses over the radio. "We're in convective nirvana!"

At 3:15, we pull off the road to wait for the arrival of the rain-wrapped hook. Rain comes lashing down in sheets, then hail the size of marbles, hammering on the car like machine gun fire. By 3:20 we find ourselves being overtaken by an absolutely magnificent prairie storm, with great, towering ramparts of darkness rising above the earth. Where light bursts through the roiling darkness, it has a distinctly greenish cast, a sure sign that there is ice in the higher elevations. That spooky greenish light, and hail, are two calling cards that often accompany tornadoes.

But we are not alone. At almost every rural intersection we come to there is a traffic jam of cars and vans, waiting for the storm. Some of these are scientists and researchers; some are police and fire department vehicles, which often act as storm spotters in small towns; and some are commercial storm-chasing tours. The net effect is of a media circus, a slow-motion riot. It's like a rock concert. And, in fact, it *is* a rock concert, except better, because these great atmospheric extravaganzas of sound and light dwarf any rock-and-roll spectacle staged by Pink Floyd or the Rolling Stones. People come here from around the world to see the show—we see a group called the Danish Severe Weather Society and a Japanese group—because although tornadoes have been seen on all continents except Antarctica, three-quarters of them occur here on the great prairies of the United States.

We stop near a crowded intersection, and a guy comes running up to our vehicle in the pouring rain, holding a video camera covered with a plastic bag.

"May I interview you? I'm doing a film," he says, in a distinctly British accent.

"No," Tony says.

There are people standing on the roofs of cars, even though this immense storm cell has now begun to crackle with lightning, as if it were coming alive. Bruce Lee later observes that when storm chasing first became popular, just after the movie *Twister* came out in 1996, people would find a safe location miles away to watch these storms. But they keep getting closer and closer, almost daring the fates to strike. Now we are out here directly in the dragon's path, and it's spitting fire and smoke as it bears down on us.

"Let's get out of this circus and head south on 24," Tim says over the radio. "We'll try to catch the cheeseburgers that are coming along behind this thing!"

"Cheeseburger" is one of Tim's almost endless supply of culinary metaphors to describe tornadoes. We drive south on Route 24, and at 4:15 stop again to watch the massing of an immense wall cloud. Something seems to be forming along its bottom edge, like a fat trunk, as if the storm were attempting to form a funnel. Above it, murky light streams down through a sort of window in the dark cloud. This formation is called an RFD cut, and it shows where the rear flank downdraft has broken open a chink in the wall of cloud—a promising development if you're looking for tornadoes. But somehow the whole structure rapidly becomes disorganized and falls apart. It's as if a fearsome army has suddenly thrown down its arms and run off.

Tony, checking GRLevel3 on his laptop, notices that the storm cell seems to be heading west. He also notices that in the confusion we've become separated from M1 and M2. Following the radar, we head back west on the interstate, through the town of Grainfield, then turn north onto another mud-gravel Kansas road that's been turned into cake batter by rain and hail. Above and slightly to the north of us, there is a whitish, diffuse cloud mass, not distinguishable as a supercell at all. A barely discernible white tongue of cloud begins to take shape. Then it narrows and begins descending. Moments later the white tongue, now more like a finger, has descended all the way down to the ground.

"Tornado on the ground!" Tony howls into the radio as we speed west, trying to find a road to cut north and follow the twister, which appears to be moving away from us toward the north at a rapid rate. We find a muddy north-south road and turn onto it, as the whitish tornado begins a fantastic series of shape shifts, first into a long, sinuous, snaking rope; then into a snaking stem with a lovely fanlike protuberance on top, like a drifting aquatic plant; then into a long, white funnel with what appears to be a humanoid mass at the top, as if it were a python that has swallowed a body; then into an almost vertical funnel disappearing into the shapeless clouds above. The tornado is almost pure white, absolutely beautiful, and quite far away, so that the net effect on the viewer is one of sheer aesthetic pleasure, without a trace of fear. One might say (quite unscientifically speaking) that the tornado has a distinctly feminine appearance. In fact, Tony later calls it the "Cindy Crawford" tornado.

Now Tony is multitasking like mad, one hand on the steering wheel, the other clutching a digital camcorder, as we mount a hill and tear off down the muddy road after Cindy Crawford. Suddenly the rear end of our ridiculous little car begins to go into a long, slow, swooping fishtail to the left.

"I got it," Tony says calmly, frantically spinning the steering wheel in the opposite direction with his right hand, trying to straighten out the front wheels without touching the brake. The back end reverses course, fishtailing out to the right.

"I got it, I got it." Now the rear fishtails to the left again, careening like a boat down the muddy hill. The vehicle is only going about 30 miles an hour, but it is not slowing down, and the sensation of careening downhill at the very edge of out of control is, to say the least, unsettling.

"I got it, I got it."

Suddenly he doesn't have it and the car goes flying over the grassy shoulder on the left side the road, becomes momentarily airborne, teeters on the brink of tipping over, then rights itself and lands with a *whump* in the muddy ditch. Tony and I let out a whoop of relief and exhilaration.

"Way to go, man! Nice driving!" I howl, grabbing him by the arm.

Even though, let's face it, we're in a ditch, at least we're not upside down. We scramble out of the car just as another storm-chasing vehicle flashes past and its occupants wave without stopping. Tony runs up the road, shifting between a digital camera and a digital video camera, as Cindy Crawford retreats into the distance. One can clearly see the whitish condensation spinning around the core and moving downward, toward the ground, like the white wraiths produced by dry ice. Unseen to the viewer, the interior of the spinning funnel is actually rotating upward, toward the sky.

Tony is totally pumped from this close encounter with disaster, and the fact that at least we've spotted a tornado for the day. Our decision to leave Denver and go east to Kansas was not entirely mistaken.

His cell phone jingles with a text message. It's another chaser, wanting to know if he's OK. "How did you know I wiped out?" He text-messages back: "Because the chaser behind you had a webcam mounted on his dashboard, providing live streaming video on his Web site." The great Tony Laubach wipeout of 2008 had been witnessed by the more than 400 people who were watching the webcam at that particular moment (on SevereStreaming.com).

Now Tony's cell phone starts to jump up and down with voice mail and text messages from other chasers who had watched him in action. The 21st century has in effect loaded up all these chase vehicles with hundreds and even thousands of other passengers, each one participating in a virtual-reality tornado chase, many of them from a great distance away, perhaps even another continent.

But we are not in virtual reality. We're in reality reality, which is muddy and inconvenient, and our very real car is still stuck in the ditch. Tony attempts to back up, but the wheels just spin out helplessly. He tries to turn the vehicle around, but it doesn't want to go. Finally he straightens the wheels and just drives forward slowly, parallel to the road, as the ditch gradually gains elevation until the road and the ditch meet and he's able to drive out. Hey, no problem.

But by now the Cindy Crawford tornado has disappeared, having dissipated gradually into the white cloud mass whence she came. We drive back into the little town of Grainfield, where we stop at a tiny local gas station beside the interstate. The power is out. An attendant at the station tells us he saw the tornado touch down right on the lawn out front, then lift off and head north. He says he heard over CB radio that the tornado knocked over a tractor trailer just down the interstate.

"Where would you go if a tornado hit?" I ask.

"Right there," the attendant says, pointing to a door ten steps from the cash register, where a narrow stairway leads down to a dim, cool storm shelter. Out here tornadoes are just a fact of life, like death or gravity.

Outside it's pouring down rain. Inside, there's no light in the men's room, so it's necessary to use the faint bluish light from my cell phone to light the way. It must be admitted: There is a certain mad glee that comes out in people when nature has run riot, the electricity is out, the phones don't work, and some brute spectacle is unfolding all around, in glory and ruination.

Now Tim's voice crackles over Tony's radio.

"M1, M2, and M3, where are you? I guess we lost radio contact with you guys because we've been chasing a tornado about three miles south of Quinter. The crazy thing got so rain-wrapped we couldn't see it, so we gave up and headed north."

"Probe, we're on I-70 heading east toward WaKeeney—looks like there's something interesting developing over there," Tony radios back.

"Yeah, Threat Net is showing a little rotation on the other side of this precip, so let's hurry on over to WaKeeney and have a taste," Tim responds.

We head east toward the small town of WaKeeney on the mostly deserted interstate, in driving rain. It's 6:20 and the sky is getting dark. In fact, we seem to be driving due east into hell, which is dead black with weird greenish highlights.

"I'm seeing a hook echo with TVS just south of WaKeeney," Tim radios to the others. "Let's just push through this P.O.S."

Incongruously, we pass a small armada of immense farm vehicles, like giant insects, creeping down the highway. Most other vehicles have pulled off the road and are stopped on the shoulder with lights blinking. Suddenly Tony erupts.

"Tornado on the ground!"

Off to the right of our vehicle, perhaps a half mile south of the interstate, where the black clouds hang down like a too-low ceiling, there is a fat, evil-looking tornado. It is moving directly toward the interstate (and us). It's mostly translucent; you can see huge objects, perhaps trees or rooftops, swirling around inside it, but it is not the solid mass of condensate that makes many tornadoes appear to be solid. Everybody in the TWISTEX team has pulled off the highway. People jump out of vehicles and then hang on to them for dear life because the edge of the mesocyclonic circulation is coming directly at us. The same tornadic wind that 20 minutes ago knocked over a tractor trailer is now clobbering the crummy tin bodies of the rental cars, Verne Carlson's old Subaru, and the geekmobile. When the tornado passes across the highway to the west of us, perhaps a quarter mile away, it takes two hands to pull the car door shut. It's as if the wind has become a living, breathing thing, a thing that does not wish us well.

Carl Young will later remark that this is the closest he has ever come to actually getting *inside* a tornado.

For Tony Laubach, this day with its multiple tornado sightings, car wreck, big disappointments, and hair-raising chases, has become the sort of day he lives for.

"Storm chasing is like this perfect thing for me," he says. "I love seeing all this crazy weather, and I love traveling around—I've probably driven every highway east of the Rockies. I love winding up in places like Valentine, Nebraska, or Seymore, Texas. Who goes to these places? Well, all these little towns and all this crazy weather, that's what I love. People ask me if I go to church, and I tell them, 'Yeah, every May and June, out west.'

"That's church to me."

7: THE "BIG KAHUNA"

LATE ONE AFTERNOON IN AUGUST 2006, ON A LONELY DESERT HIGHWAY SOUTHEAST of the tiny town of Las Vegas, New Mexico (not *the* Las Vegas), Tim Samaras pulled his instrument-laden chase vehicle off the highway onto a prominent knoll and parked. In the vehicle with him were meteorologist Carl Young and German photographer Carsten Peter, who watched as Tim maneuvered his truck so that its rear end was facing directly northwest, toward the gathering storm. The task was a little difficult because attached to the rear end of the vehicle was a gray metal trailer about the size of a low-budget U-Haul. Once the vehicle was properly positioned, Tim jumped out, ran back, and dropped open the trailer's tailgate door.

In the dimness inside the trailer, the faint light caught a collection of spectral eyes, like owls' eyes, peering out at the storm. The "eyes" were the lenses of a series of shelf-mounted high-speed cameras and other photographic equipment. Tim climbed into the trailer to make adjustments to all this gear as, in golden late afternoon light, the atmosphere over the desert floor boiled up into a tremendous summer thunderstorm, riven with lightning.

Much of the American Southwest is desert or dry grassland, but during the months of July through September the "monsoon season" brings frequent warm summer rain, thunder, and lightning. During August and September, Tim and his team range out into eastern Colorado, New Mexico, west Texas, and Arizona. They're not looking for tornadoes (this late in the summer, tornado season has long passed). They are looking, instead, for another atmospheric phenomenon that is equally dangerous and equally mysterious: lightning.

Tim Samaras has been fascinated by lightning for almost as long as he's been fascinated by tornadoes. When he was a child and a summer thunderstorm came booming and flashing over the family house in Denver, his mother used to draw the blinds in his bedroom so he wouldn't be frightened. But Tim always pulled the blinds open again. He wanted to see the whole spectacle with his own eyes.

He also discovered that lightning strikes create powerful radio waves, or "lightning signatures," that travel around the world by bouncing off the ionosphere (a reflective atmospheric layer wrapped around Earth at an elevation of 50 to 250 miles). In the evening, he discovered, you can tune an AM radio between stations and hear the beautiful dispersion ringing of the static from lightning strikes as it bounces back and forth between Earth and the ionosphere. Usually referred to as static crashes, these eerie echoes from the edge of space could sometimes sound like hundreds of tiny bells ringing at once. As a kid, Tim was entranced by the notion that these sounds were being transmitted around the rim of the world. He used to lie in bed at home in Colorado, with his AM radio tuned between stations, listening to thunderstorms in the Midwest.

When he relates this story to a visitor, he lifts his eyebrows sheepishly. "Yeah, I know," he says. "What a geek!"

In his early days of tornado chasing, Tim made use of this boyhood pastime. In those days, there was no wireless Internet, no cell phones, no access to radar—in fact, there was almost no real-time data at all to help chasers predict tornadoes. Instead, what Tim Samaras had was his car's AM radio, tuned not to the local rockabilly station but to the music of the atmosphere. By listening for static crashes and judging how loud they were, he could get some general idea of how far away the nearest storm cell might be.

There were almost always static crashes on the radio. People tend to think of a lightning strike as a random, anomalous event, which is why statisticians have taken the trouble to calculate one's risk of being struck by lightning (in the United States, your chance of getting struck in any given year is about 1 in 700,000). But if one were to spiral up into near Earth orbit—in the space shuttle, say—one would see a planet alive with the near-constant flicker of lightning. At any given time, there are about 2,000 thunderstorms in progress, which collectively produce around a hundred cloud-to-ground lightning strikes per second, or more than eight million lightning bolts a day.

From that Earth-girdling perspective, you could almost say that the *lack* of lightning is an anomalous condition. Lightning is not just an occasional occurrence, but the natural state of things—as natural as a heartbeat.

In fact, lab experiments have shown that electrical discharges in what is thought to be the constituents of the primordial atmosphere can create complex molecules necessary for life. It's possible that it was a lightning strike (or strikes) that touched off the cascade of events that led to human civilization.

It's not precisely clear what the relationship is between tornadoes and lightning. Some recent scientific studies have suggested that lightning activity actually *decreases* in and around tornadoes. Tim says he has not observed this personally. One thing that's certain, though: Lightning doesn't *cause* tornadoes, which is what was erroneously believed until the late 19th century. Since a certain amiable printer from Philadelphia flew a kite into a storm one summer afternoon in

1752, it had been known that lightning is an electrical phenomenon. (Luckily, Benjamin Franklin's famous kite was not actually *struck* by lighting; it "merely" flew into an electrically charged cloud, which sent a current pulsing down the wet kite string.)

And it had been observed that fairly often (though not always) there was lightning in the big frontal storms often associated with tornadoes. It was also known that electricity can sometimes cause things to rotate. Hence, early on it was thought that lightning did cause tornadoes.

On May 30, 1879, a tornado tore through Delphos, Kansas, whereupon John H. Tice, the self-proclaimed "weather prophet of St. Louis," opined that the town's electric circuitry had drawn down the storm. "Railroad and telegraph lines obey the laws of induction and give rise to the necessary electrical charges to produce storms," he explained, pointing out that *all* weather is electrical in nature. There was just one problem: Delphos, Kansas, *had* no telegraph lines, and no railroad either. Town boozers just claimed to have them. This information did not dissuade "Professor" Tice or other, more qualified, meteorological researchers, who continued to believe that atmospheric electricity might have something to do with tornado formation as late as the 1960s. (Although most tornado scientists now consider this link entirely disproved, there are still unexplained oddities about tornadoes—high-frequency radio waves and eyewitness reports of strange lights in nighttime tornadoes—that "seem to suggest some unknown electrical processes taking place within some tornadoes," according to electrical engineer William Beaty.)

While the debate over the relationship between tornadoes and lightning continues, one thing is for sure: Lightning is really, really fast. In fact, one might say that the mysteries of lightning are hidden inside almost unimaginably slender microslices of time—milliseconds (thousandths of a second), microseconds (millionths of a second), and even tens of nanoseconds (billionths of a second). The return stroke of a lightning strike travels about one-third the speed of light. These phenomena are so brief that no one has ever filmed them with a camera fast enough that the film could be replayed in slow motion to reveal what is actually occurring. It's like instant replay in a football game, Tim likes to explain to kids: There are things about lightning that have never been witnessed by humans, the way a fumble can be witnessed in slo-mo during a Redskins game.

And there's nothing that gets Tim more excited than attempting to see, and to photograph, that which has never been seen or photographed before.

The normal home camcorder has a framing rate of about 30 frames per second, he explains. That's far too slow to capture anything of scientific value about lightning (except for the occasional stroke-of-luck shot that occasionally appears on the Internet). But what if you used the sort of highly specialized, super-high-speed digital cameras commonly used to photograph explosions, which Tim had been doing as part of his job for decades? Tim was, in fact, the first person to hit on the idea of using a high-speed, high-resolution digital camera, called a Phantom V7.3, to attempt to photograph lightning. A Phantom camera records at speeds up to an incredible 15,000 frames per second.

These cameras, manufactured by a New Jersey company called Vision Research, cost $80,000 to $100,000 and are generally used in industry or manufacturing to study any phenomenon too fast to be seen by the human eye, such as bombs, bullets, or high-speed machinery production. For instance, Tim says, "If you've got a machine that stamps out 100 beer cans per second, and you've got some sort of glitch you can't see or figure out, an engineer can use one of these high-speed cameras to find what the heck is wrong."

Phantom cameras perform beautifully on these mundane industrial tasks. But nobody had ever thought to lift a Phantom camera from the mundane (photographing a beer can machine) to the celestial (imaging a bolt of lightning).

Of course, you have to *find* a bolt of lightning and have the camera properly positioned and aimed in order to capture it. Some meteorological laboratories have high-speed cameras, but researchers have to wait until a thunderstorm comes to *them,* like a wallflower waiting to be chosen at the dance. Instead, Tim designed a sort of mobile lightning lab that he could take out in search of storms—a small trailer, with two Phantom cameras mounted on camera shelves, with special suspension to minimize road vibration, as well as a dash cam mounted on the dashboard.

Now, as the crackling electrical storm rolled in over Las Vegas, New Mexico, during that summer of 2006, Tim noticed a couple of things about the whole situation that got him genuinely excited. First, they had the trailer in position, angled directly at the storm, and the cameras were up and running. Second, the storm was producing lightning "like crazy." And third, the storm was moving along at a very leisurely pace, perhaps five knots. It was like one of those rare tornadoes that, like a temptress, seems to deliberately slow and momentarily reveal itself. Huddled inside the trailer out of the rain, Tim let the Phantoms rip.

Once the storm had passed, Tim, Carl, and Carsten reviewed the video of what had just happened. When they saw what they'd gotten on memory, Tim said later, "I was totally blown away." It was the first time in human history that a high-speed, high-resolution camera had captured the whole story of a lightning strike.

In the video (which so far has been shown only at scientific conferences), the first thing a viewer notices is a tractor trailer in the foreground, flying down the interstate at about 70 miles an hour. But because of the Phantoms' speed, the truck appears to have come to a complete stop. High above, out of the dark clouds, there is what appears to be a falling spark, like a rogue spark popping out of a campfire. Then another spark, then another, until there appears to be a kind of firefall of sparks, like the sparks flying off a grinding wheel, each one relatively random but all of them trending downward. Then one of these sparks appears to branch into stair steps on its way down—the familiar zigzagging pattern of a lightning bolt, known as a stepped leader. Then there is a series of these relatively faint stepped leaders, until one appears to make it all the way to the ground and there's a tremendous, instantaneous, pulsing bolt—vibrating almost like a strobe light—which engulfs the frame with light. This return stroke is a river of electrons pouring back up the channel burned by the stepped leader.

"The front edge of the stepped leader sparkles on its way down—I just find that astounding," Tim says, his voice still tinged with awe at what the video showed. The sparkling is fairly rapid, at a rate of about 15 kilohertz (15,000 cycles per second), he says.

"For years, lightning was shown in textbooks as simple cartoons, but these images are likely to be used in future textbooks, showing that the whole process is a lot more complex."

Why does the stepped leader sparkle?

"I really don't know," he says. "There are people who spend their whole lives working on plasma theory, and some have suggested it might be the ionization path of the stepped leader, flaring as it hits oxygen molecules, or some other phenomenon."

After Tim "let the cat out of the bag" with this breathtaking 2006 high-speed video, other people went out and rented or borrowed Phantom cameras and began producing footage of these same phenomena, some of which were posted for all the world to see on YouTube. Scientific papers were written describing the firefall of sparks as "recoil streamers." Even so, Tim Samaras did not try to "claim" this discovery. To him it would be like trying to plant a flag in a lightning bolt.

"People ask me, 'Doesn't that bother you at all, that other people are out there writing papers about this?' Well, not really. I just love the challenge of trying to acquire knowledge, trying to examine things for the first time. Also, I'm kind of in a unique position because I'm not in academia, so there's not all this pressure on me to claim discoveries and publish papers in peer-reviewed journals and so on."

The joy of pure research was the primary thing that motivated Benjamin Franklin, who was the first to surmise (at least approximately) what causes lightning in the first place. In a 1749 letter, he suggested that particles of water in a thundercloud become electrically charged by being jostled around in the wind; lightning is simply the discharge of this pent-up electrical force. This explanation, though crude, is not so terribly different from what is now understood about the origins of lightning.

Lightning occurs because of the tendency for electrical charges to separate in a thundercloud, with positive electric charges building up in the upper ramparts of the cloud, and negative charges accumulating in its lower elevations. Air is a fairly good insulator, so the positive and negative fields remain separated, at least temporarily. In effect, the great piling cumulonimbus storm towers of summer are like giant capacitors in the sky (a capacitor being an electrical device which builds up and then stores opposing electrical charges by keeping them insulated from one another).

But charge separation in an electrical field is inherently unstable, and it doesn't last very long. Lightning is essentially a giant electrical spark that neutralizes two opposing electrical fields, akin to the static electricity that jumps from your finger to a doorknob when you walk across a thick carpet.

If the negative field is in the cloud and the positive field is in an oak tree, the flash is called cloud-to-ground lightning (what chasers refer to as CG). If it's within or between clouds (actually, the most common kind of lightning) it's referred to as intracloud or cloud-to-cloud lightning.

Cloud-to-ground lightning southeast of Blackwell, Oklahoma, May 2008. Lightning strikes occurred every couple of seconds at the storm's peak.

Like a tornado, a lightning bolt is a cavalcade of curiosities. A lightning bolt can often reach over five miles—one bolt, near Dallas, was actually measured at more than a hundred miles. It can raise the temperature of the air to as much as 50,000 degrees Fahrenheit, which is five times hotter than the surface of the sun. It's so hot it can fuse sand into glass. Lightning is so hot, in fact, that it "explodes" the air, causing it to expand instantly, which creates a compression wave, or shock wave, that travels outward in all directions. We hear it as thunder. One could think of the rumbling sound of summer thunder as an acoustic "signature" of a particular lightning flash—an auditory representation of the shape, size, and intensity of that explosion. The flash of the explosion is often brighter than ten million

100-watt lightbulbs. And (just to point out how dangerous Ben Franklin's experiment was, as children's science books inevitably do) a lightning bolt typically contains hundreds of millions of volts of electricity, and sometimes a billion or more.

In a cloud-to-ground lightning stroke, the initial discharge or downward traveling spark (or stepped leader) usually starts at the base of the cloud. How the negatively charged stepped leader actually works is not known. But what appears to happen is that a low-luminosity spark finds its way from cloud to ground in a series of jumps, each about 150 feet long. It branches and forks down through the irregularly dispersed ionized air mass as it descends. This progress of the step leader is actually relatively slow (at least, slow in the high-speed world of lightning). Each leader step may last less than a few milliseconds to several tens of milliseconds (thousandths of a second). The average velocity of the stepped leader is about 93 miles a second, so that the entire trip from cloud to ground may take 20 thousandths of a second. This whole initial phase involves only a relatively small electrical charge and is practically invisible, especially compared with the pyrotechnics soon to follow.

As the negatively charged stepped leader approaches the ground, it draws out a positive charge, or positive streamer (sometimes called a dart leader), from objects below (since opposite charges attract each other). On the few high-speed photographs that have captured it, positive streamers show up as spooky, bluish tentacles of light emerging from the top of a house or a tree. The descending, negatively charged stepped leader meets the ascending, positively charged positive streamer a few tens of yards above the ground, in the attachment process.

Once this magic moment occurs, and the circuit is completed, the stepped leader's negative charge crashes violently into the ground, with an enormous, brilliant flash. Then the blinding-bright luminosity flows back up the channel into the cloud. So, though your eye can't follow it, the brilliant flash of lightning, or return stroke, actually goes *up* rather than *down*. The electric circuit flows back up the leader channel at somewhere between 20,000 and 60,000 miles per second, a fraction of the speed of light.

You don't need a degree in physics to understand at least one thing about lightning: It can kill you. In fact, lightning causes more direct deaths than any other weather phenomenon. (Though tornadoes inflict much more physical damage, they cause only about half as many deaths as lightning.)

Tim tells the story of another storm-chasing trip, also in the company of Carl Young and Carsten Peter. They had pursued a ferocious storm from New Mexico all the way across the border into the Texas Panhandle, by which time the Storm Prediction Center at the National Weather Service had issued a tornado watch. Suddenly the trio of storm chasers found themselves "getting the snot beat out of us" by softball-size hail, Tim remembers. Still, they parked their mobile lightning lab beside the road, opened the trailer door, and tried to get the Phantom cameras up and running. But every five or ten minutes, they'd have to pack up and move because the storm, exploding with lightning, would overtake them.

"The flash-to-bang ratio was practically instantaneous," Tim says—not a good sign, because it meant that the lightning was striking so close it was almost tapping at their back door.

Nevertheless, the intrepid Carsten went scampering up a nearby hill and began setting up his metal camera tripod in the teeth of the storm, as though goading the gods of lightning. He was, as usual, giddily engrossed in his work when suddenly Tim saw a lightning bolt slam into a field not 50 feet away from him; moments later he could see a wraith of smoke rising up out of the ground.

"I never saw a German jump that high!" Tim recalled, as he watched the beloved Deutschlander hightail it down the hill toward the truck.

Tim Samaras's lack of concern that other researchers might "scoop" his high-speed lightning pictures has to do partly with his high-mindedness and devotion to pure research. But it also has to do with the fact that he has one other little trick up his sleeve. Actually, it's a very *big* trick, one that weighs 1,600 pounds and stands six feet tall. It's the primary reason for that "mobile lightning lab" he drags behind his truck.

Tim calls it, simply, the big kahuna—the camera to end all cameras, a machine that will make the Phantom look about as up-to-date as a daguerreotype of Abe Lincoln. The big kahuna is the fastest, highest-resolution camera in the world. Sure, the Phantom can expose a mind-boggling 15,000 frames per second—but this thing is a hundred times faster, exposing an almost inconceivable 1.5 *million* frames a second. It's so fast, so far beyond the reach of the human senses, that it seems nearly capable of stopping time itself. It's a machine that will absolutely smoke the competition. It fact, there won't *be* any competition. (Or, as Tim puts it more modestly, "I think I'm the only nut out there doing this.")

The story of Tim's association with this amazing camera began back in 1985, when he was employed by the University of Denver Research Institute, where he'd gotten a job straight out of high school. Among other things, DRI did ballistic impact, explosive firing, and hyper-velocity impact testing, which is a fancy way of saying they blew things up and photographed them at extremely high speeds. As part of this research, the company got access to an extraordinary, highly specialized piece of equipment owned by the U.S. government, called a Beckman-Whitley Model 192 ultra-high-speed camera.

"This thing was an absolute monster," Tim recalls. "It weighed almost a ton. It was a Cold War relic, designed during the mid-1960s to photograph nuclear bombs going off."

In 1985, in preparation for an extremely large, aboveground non-nuclear weapons test, the camera was shipped from California to an explosives bunker in the desert at White Sands Missile Base, near Socorro, New Mexico. Though Tim had never even seen this mega-camera before, he was dispatched to White Sands to make sure it was operational. (To clarify this point: At the time, Tim was all of 28 years old, and not only did he lack an engineering degree, he had never even been to college. But "I had done lots of crazy stuff and had proved myself at the Research Institute," so he was trusted to get the camera up and running.)

Actually, the Beckman-Whitley was not really a camera at all. It was *82* cameras, arranged in a ring, like a big wheel about six feet in diameter, all pointing toward the center. (Since this predated the digital era, these cameras used photographic film.) At the hub of the wheel was a small three-sided mirror, about three-quarters of an inch wide and seven-eighths of an inch long. This was the primary moving part. Once the composite camera ground into action, this tiny optical mirror began spinning, flashing the image coming in through the lens to each camera in turn. Driving the spinning mirror was a helium-powered turbine that could reach speeds of up to 4,500 revolutions per second. That was so fast that the camera could be run at top speed for only 30 seconds or so. If it overheated, or if the mirror came loose and ricocheted through the device like a bullet, or it threw a bearing, the camera could be destroyed.

Tim succeeded in getting the camera functioning flawlessly and "it worked great" for the explosives test, producing reams of good data. (The test itself was the largest conventional explosives test in world history, equivalent to detonating 600 tons of ammonium nitrate in order to simulate a nuclear bomb.) After the big explosives test, Tim loaded the camera onto a trailer and shipped it back to Denver, where it was used on a series of smaller explosives tests. Tim became intimately familiar with the quirks and capacities of the Beckman-Whitley—in fact, as time went on, he became the *only* person at the institute who really knew how to operate the beast.

There were times, he admits, when he was using the camera at DRI's explosives test site outside Denver and thunderstorms would go rumbling by overhead and he was strongly tempted to raise the big camera's all-seeing eye up to the heavens to see if he could follow his dream and capture a lightning strike. But the device was government property, and taking lightning pictures with Uncle Sam's camera would have produced a detailed photographic record of his transgressions—like one of those security cameras at your local ATM, but multiplied by millions.

Fast-forward 20 years. In 2005, long after Tim had gone on to work for an Albuquerque-based engineering firm, Applied Research Associates, he got a call from an old friend who was still working at DRI.

"Hey, I thought you might like to know that we're selling off a lot of old government surplus stuff, and that old Beckman-Whitley camera is going to be on the block. You could probably get it for next to nothing, because nobody around here even knows how to run it."

Tim went home and asked his wife, Kathy, if she'd mind if he bid on some government surplus equipment.

"My poor wife knew I had a long history of collecting strange electronic junk at auctions, so when I asked her if I could buy this old camera, it didn't really faze her. She said, 'Well, whatever, you know.' Just another day of living with Tim."

So Tim put in a modest bid for the camera and won it, there being no other bidders standing in line. Unfortunately, when Tim came home with this 1,600-pound behemoth in the back of a truck, Kathy changed her tune. *"You're not bringing that thing in this house!"* she said, standing in the doorway with her arms crossed. She had been quite forgiving over the years, after all: The family garage was crammed full of electronic gadgets and gizmos, and towering over their ordinary middle-class home in an ordinary middle-class neighborhood was a 105-foot ham radio tower, so high they had had to get a special variance from the local zoning commission.

Tim dutifully found another place to stash his new toy—a rented storage locker—and there it sat gathering dust for over a year before he got a chance to begin trying to bring the camera back to life. There were a couple of challenges to adapting the camera to trying to photograph lightning. For one thing, the lens had such a tight field of view that it required finding a very prolific storm with a lot of lightning occurring in a relatively small area (not easy). The other challenge was film—82 cameras exposing more than a million frames per second results in a lot of exposed film (and more than likely, most of it worthless). Tim took the big camera out for about a week during the monsoon season of 2006 but did not succeed in overcoming all these problems.

Then he and technicians from the National Geographic Society, who had generously funded his work and were extremely interested in the camera, decided to convert the beast to digital. During 2007 and 2008, the big camera was taken apart, and everything having to do with film was unbolted and replaced with digital technology and electronics. Now 82 CCD imagers (essentially the sensors in a digital camera) replaced the old film cameras. The digital cameras will produce high-resolution images (11 megapixels, where 8 megapixels is the norm). The nearest competitor, the Cordin Company, in Salt Lake City, is working on a high-speed camera with only one megapixel of resolution. The plan is take the big camera out into the field during the summer of 2009.

What, exactly, is Tim hoping to see?

"The thing that I am really fascinated by—the thing that nobody has ever seen or photographed before, is the 'attachment' process—how lightning actually connects with objects on the ground," Tim says. "I've seen lightning strike the ground instead of a 50-foot power pole nearby. Why does it seem to 'select' targets like that? This is one of the secrets we're hoping to uncover. Not knowing: That's what really gets me going."

8: THE MAN WHO RODE THE THUNDER

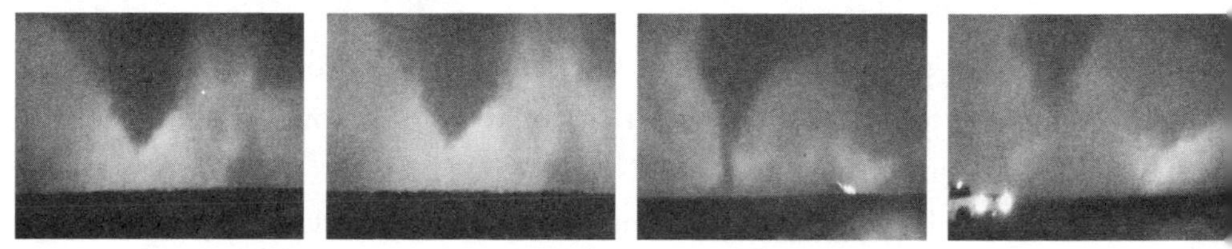

ON THE AFTERNOON OF JULY 26, 1959, A SUNNY SUNDAY, MARINE LT. COL. William Rankin roared off the runway at the Naval Air Station at South Weymouth, Massachusetts, near Boston. The plane he was flying that day was magnificent: A sleek, swept-wing F8U Crusader, a supersonic jet. Right behind him, another F8U took off, piloted by a junior airman named Lt. Herb Nolan. Colonel Rankin, 39, was taking Nolan up for a "check out" flight in the Crusader—what looked to be a straight and level, somewhat dull, one-hour trip down to Beaufort, South Carolina.

In all his years of flying, Rankin had never lost the boyish thrill he got when he slammed the throttle into the afterburner notch and the plane broke free of the earth, climbing to altitude at astounding speed.

"Tiger One, this is Tiger Two coming aboard, on your starboard side," Nolan radioed to Rankin, as he pulled his plane alongside.

"Roger," Rankin radioed back. "Let's go! Full throttle—destination Beaufort. Boy, this is living!"

The weather was brilliantly clear until Rankin began approaching Norfolk, Virginia. There, abruptly, immense, dark storm towers began piling up to more than 45,000 feet. As a student aviator, Rankin had constantly been reminded of the dangers of flying through thunderstorms—just recently, a young, three-man civilian test crew had tried to penetrate a thunderstorm and been forced to bail out; all three were killed.

One great thing about the F8U, though: You could usually just fly over the *top* of a thunderstorm. Rankin now throttled up the jet, clipping through the wispy tops of the cloud towers, breaking 47,000 feet and climbing.

"Tiger Two, this is Tiger One," Rankin radioed Nolan. "It looks as if Norfolk is catching hell. That's quite a storm down there."

"Sure is, Colonel," said Nolan.

Suddenly Rankin heard an ominous *thump* followed by a rumbling sound underneath his seat. Then a warning light went on in the instrument panel. The light was about the size of a quarter, traffic-light red, and it said: FIRE. Of all the warning lights on the instrument deck, it was the one that unnerved pilots most. Rankin eased back on the throttle, and inexplicably, the light went out. Then, a couple of seconds later, momentary relief turned into full-blown alarm: The rpm indicator suddenly dropped from 90 all the way down to zero. For whatever reason, the engine had just seized up and died. He yanked on the handle to start the backup power package, a sort of onboard generator that would reactivate the instruments, the radio, and the hydraulic steering; but to his disbelief, the handle came off in his hand. Rankin was now essentially riding a tin can going 625 miles an hour, nine miles above Earth.

"In a crippled high-performance, high-altitude aircraft, sweeping up past 47,000 feet, only a few thousand feet from the dividing line between earth's atmosphere and true space, you don't waste time on personal thoughts," he wrote later in a book called *The Man Who Rode the Thunder*. "You're much too busy fighting for your life."

He had only a matter of seconds to decide what to do. He had never heard of anyone ejecting from an aircraft this high up—and he was not even wearing a pressure suit, meant to protect a pilot's life at high altitude. He had on nothing but an ordinary summer-weight flight suit, gloves, helmet, and shoes, as if he were about to go for a stroll on the boulevard.

Perhaps worst of all, below him lay the towering thunderstorm that he would later describe as "one of the most violent storms ever to strike the East Coast." Although the term was not widely used at that time, the storm was very likely what we would now call a supercell—one of the rarest, most extreme, and most dangerous kinds of storms, one whose intense updrafts and mesocyclonic rotation—an area of rotation inside the larger storm—often produce tornadoes.

Supercell thunderstorm A thunderstorm with an organized, consistent, rotating updraft capable of creating tornadoes, downbursts, and large hail

Hail Balls of ice produced from a cumulonimbus cloud at least $^{2}/_{8}$ of an inch (pea size) in diameter, but growing as large as a grapefruit

As Rankin's dead jet vaulted closer and closer to the black edge of space, he realized he had no choice. Staying in the plane would be suicide. If he couldn't get the engine relit and the jet ran out of airspeed, it might tip over into a wild, out-of-control spin; or it might suddenly "tuck" and begin to drop at supersonic speed. If he was going to bail out, he needed to do it now. He reached up behind him, gripped the ejection seat handles, and pulled.

And that was how Lt. Col. William Rankin became the first person—very likely the *only* person—ever to descend by parachute through the center of what was most likely a supercell thunderstorm.

It was actually a series of thunderstorm-related airplane crashes in the mid-1940s that led to the first organized scientific study of thunderstorms, which led in turn to the modern study of tornadoes.

In the summer of 1946, University of Chicago meteorologist Horace Byers and his graduate student Roscoe Braham began studying thunderstorms in Florida, because Florida has more thunderstorms than anywhere else in North America. This work led to what became known as the Thunderstorm Project, the first organized scientific attempt to penetrate the mysteries of these storms using radar, instrumented weather balloons, and airplanes. The Thunderstorm Project was a cooperative undertaking by four government agencies—the U.S. Weather Bureau, Army Air Force, Navy, and National Advisory Committee for Aeronautics, which later became NASA—with Byers and his colleagues at the University of Chicago analyzing the data. Based in central Florida, near what would later become Walt Disney World, daring pilots would take their aircraft aloft and deliberately fly into the storm's core, letting the winds "take" the plane wherever they wished, while the pilots attempted to estimate the intensity of updrafts and downdrafts inside the storm. Though they risked their lives, none of these "top guns" was ever killed.

One of Byers and Braham's main findings was that thunderstorms go through well-defined life cycles, just like living things. They used the biological term "cell" to describe these storms, and divided the storm cell's life into three stages: cumulus, mature, and dissipating.

The cumulus stage refers to the familiar towering cauliflower clouds of a humid summer afternoon. These whitish upwellings can rise up to 35,000 or even 50,000 feet. The interior mechanics of the cumulus stage are fairly simple: The whole formation consists of updrafts, or upward-moving, warm, wet air currents. These updrafts can rise at a rate of five to twenty miles an hour or so, which may not sound like much, but it's enough to cause mild to severe turbulence if an aircraft tries to fly through them.

The second, or mature, stage is the most violent and dangerous stage of the thunderstorm's life cycle. The great humid updrafts rise to enormous heights, eventually reaching elevations high enough that they cool, condense, and fall as rain (or hail).

This action creates downdrafts, or downward-moving air. When the cool, rainy downdraft collides with the ground, it spreads outward like a powerful wave, forming what's known as a gust front, which may include violent, damaging winds.

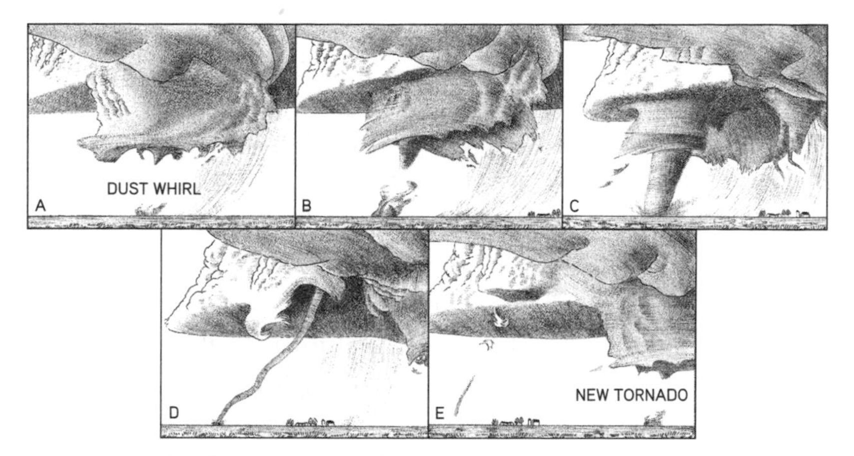

Dave Hoadley's depiction of the life cycle of a tornado: (a) During the cumulus *stage, a wall cloud forms from towering cumulus clouds; (b) a funnel descends; (c) the tornado moves into* mature *stage; (d) at the* dissipating *stage, the tornado becomes ropelike; (e) in the tornado hand-off, a new one forms from the old one.*

Eventually, in the dissipating stage, precipitation and downdrafts cool off and thus weaken the warm updrafts that powered the storm in the first place. The gust front moves away from the storm, cutting off the inflow that is fueling the storm cell. Gradually the storm dissipates, grows more disorganized, and dies.

But sometimes, for reasons that are still not fully understood, the storm story does not stop there. Under certain circumstances, a mature thunderstorm cell continues to develop. Its updrafts grow more and more powerful. Its storm tower piles higher and higher into the atmosphere. It grows ever more violent, organized, and dangerous. And, most ominous of all, its core begins to rotate. That's when it officially becomes a supercell—a monstrous storm with an immense, powerful, long-lived, rotating updraft that can sustain the storm in its most violent stage for hours. It often produces huge hail, but most alarming of all, tornadoes—sometimes more than one.

The term "supercell" was coined by British meteorologist Keith Browning (the word first appeared in print in 1963), who noticed that these super storms tended to move to the right, quite a different direction from an ordinary thunderstorm. Browning's definition of a supercell was a large-scale, long-lived, right-moving, damaging thunderstorm that persists for an hour or more.

Supercells tend to have certain recognizable features. In a supercell cloud, the updrafts are so powerful they draw warm, wet air up as high as the tropopause, the boundary that separates our atmosphere (the troposphere) from the thin, frigid layer above it (the stratosphere). The elevation of the tropopause varies, but it's around seven or eight miles high in the northern latitudes. These currents can have so much energy they burst through or "overshoot" the tropopause, sometimes by as much as a couple of miles, creating a dome- or geyserlike formation known as an overshooting top. Viewed from a weather satellite high above the supercell, this spreading cloud mass looks like pancake batter poured into a frying pan. In the middle of the pancake puddle, the overshooting top casts a distinct shadow on the cloud top in late afternoon light.

GLOSSARY

Knuckles A slang term for lumpy protrusions on the edge or underside of the anvil in a thunderstorm

Wall cloud A localized, abrupt lowering from a cloud base, which may rotate and is often an indicator that a tornado may form

Tail cloud A horizontal cloud that looks like a beaver's tail, often seen extending to the north or northeast from a wall cloud

Thermals Small, rising air currents produced from Earth's surface when heated, and the source of low-level turbulence for aircraft

Up that high, where the air is thin and the temperature is far below zero, vapor turns to ice crystals. Once rising air reaches its maximum height, it begins sinking back down and spreads out just below the tropopause, forming the shape of a blacksmith's anvil. The leading edge of this formation often has hard edges, known as knuckles. Storm chasers love knuckles because the crisper and cleaner the edges, the more powerful the updraft, the more likely the storm system will produce tornadoes. The updrafts billow up at the forward edge of a storm, and the cooler downdrafts, a kind of "exhalation," descend from the rear, trailing edge of the storm.

The bottom edge of the supercell cloud system appears as a dark, flat ceiling—a sort of horizon line floating above the actual horizon. The forward edge of the supercell tends to be completely free of precipitation. In the center of the supercell, hanging from the cloud ceiling like a pendulous udder, is a formation known as the wall cloud. The wall cloud is usually about two miles in diameter; it marks the location of the supercell's strongest updraft (in fact, the wall cloud is created by the updraft). Surrounding the pendulous wall cloud is a ringlike collar cloud; both of these structures are rotating, though often this is not visible.

The wall cloud is also the place most likely to produce tornadoes. Often, trailing out behind the wall cloud is a protuberance that looks so much like a long, horizontal tail it's known as a tail cloud.

Of course, supercells are rarely as well formed or clearly delineated as this description suggests. Often they lack some or even most of these features; or they may have features different from what a textbook might depict. Storm chasers will sometimes say that a storm tower going up looks like "a bomb going off." And actually, Tim Samaras says, there is more than a passing correspondence between a supercell and a mushroom cloud. "In an atomic explosion, you get superheated air rising, and when these thermals hit a certain layer of the atmosphere, they spread out, forming an anvil shape quite similar to a supercell thunderstorm," he says. The thermals in a supercell rise at an extremely rapid rate, similar to the velocity of rising thermonuclear clouds after the fireball cools.

But the mere description of a supercell does not do justice to the experience of actually witnessing one of these grand atmospheric spectacles.

"If you ever get a chance to see a fully formed supercell, all by itself, towering up into the sky, you will understand why we do what we do," Tim says. "It's hard to capture the experience of seeing a supercell on TV or in print, because the way you experience it out there in the field comes at you through all your senses. Its visual appearance is absolutely awe inspiring, but there is also the smell of wet wheat fields and freshly plowed dirt; the rain in your face; the sound of the wind in the power lines; and visually, what you see is not two-dimensional, it's 360 degrees, all around you. This is what chasers live for."

Chasing for tens of thousands of miles every summer, repeatedly putting one's life at risk, contending with soaring gas prices, bad food, cheap motels, separation from family—no matter. Once you see one, it is a transcendent moment that makes all the rest of the inconveniences seem about as important as pocket lint.

On the wall of the living room in Tim's house in Denver, there's an enormous photograph of a supercell in the eerie shape that is sometimes described as "the mother ship." Because that's what it looks like: an immense, multilayered flying saucer, vaguely reminiscent of the starship *Enterprise,* appearing to be settling down to Earth either to pick up passengers or to drop them off. The picture was taken on May 29, 2001, near Childress, Texas—one of those lucky days that Tim lives for.

The first thing Colonel Rankin felt after the ejection seat fired and blasted him out into the atmosphere at the top of the thunderstorm was the shock of incredibly cold air—in a flash, he had gone from a cockpit temperature of 75 degrees to nearly 70 degrees *below* zero. He was tumbling through the air at several hundred miles an hour, and every exposed part of his body—face, neck, wrists, ankles—burned like fire from the cold. Meanwhile, the pain of explosive decompression was indescribable—he looked down to see his stomach so distended it looked like he was pregnant. "Then, seconds later, the burning sensation turned to a blessed numbness," he recalled in his book.

As he tumbled down into the top layer of the storm cell, he realized his oxygen mask was flapping wildly against his face, so he grabbed it and forced it onto his face. At this altitude—he had ejected at almost twice the height of Mount Everest—the air was so thin it would quickly lead to brain damage without bottled oxygen. He struggled to stay conscious, at least until he got down to 10,000 feet, where his parachute would open automatically via a barometric pressure sensor.

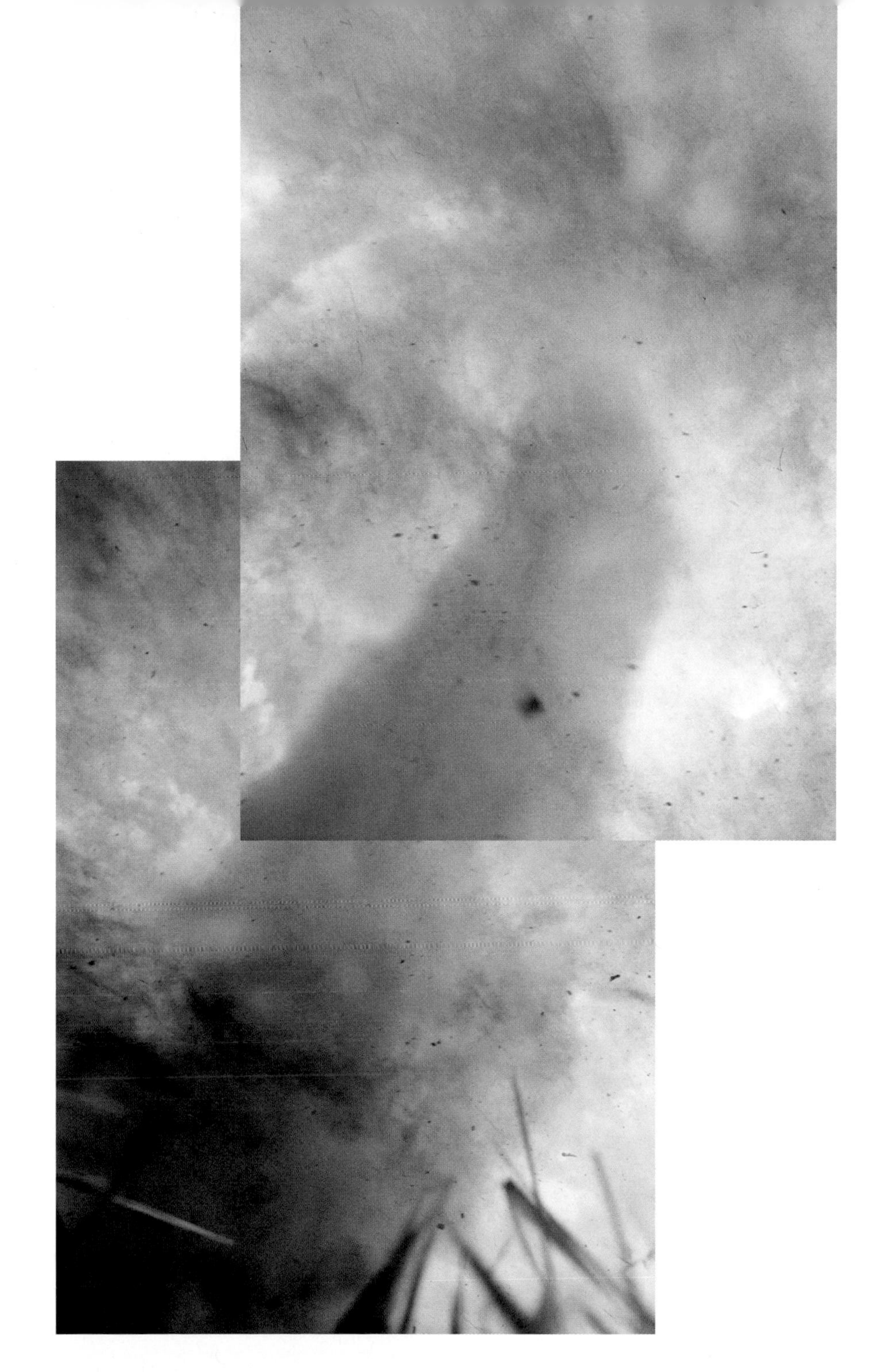

The closest images ever taken of a tornado funnel. Seconds later, the tornado tossed the camera, called the Tinman, hundreds of yards across a field.

TOP: *Tim Samaras dashes back to his chase vehicle after deploying an HITPR probe, which will measure barometric pressure.*

BOTTOM LEFT: *A photographer captures a striking lightning bolt.*

BOTTOM RIGHT: *A line of chase vehicles stands ready for action as Pat Porter videotapes the approaching tornado.*

RIGHT: *One of the 67 tornadoes to occur on June 24, 2003, on the Great Plains, an F3 tornado races across the South Dakota prairie.*

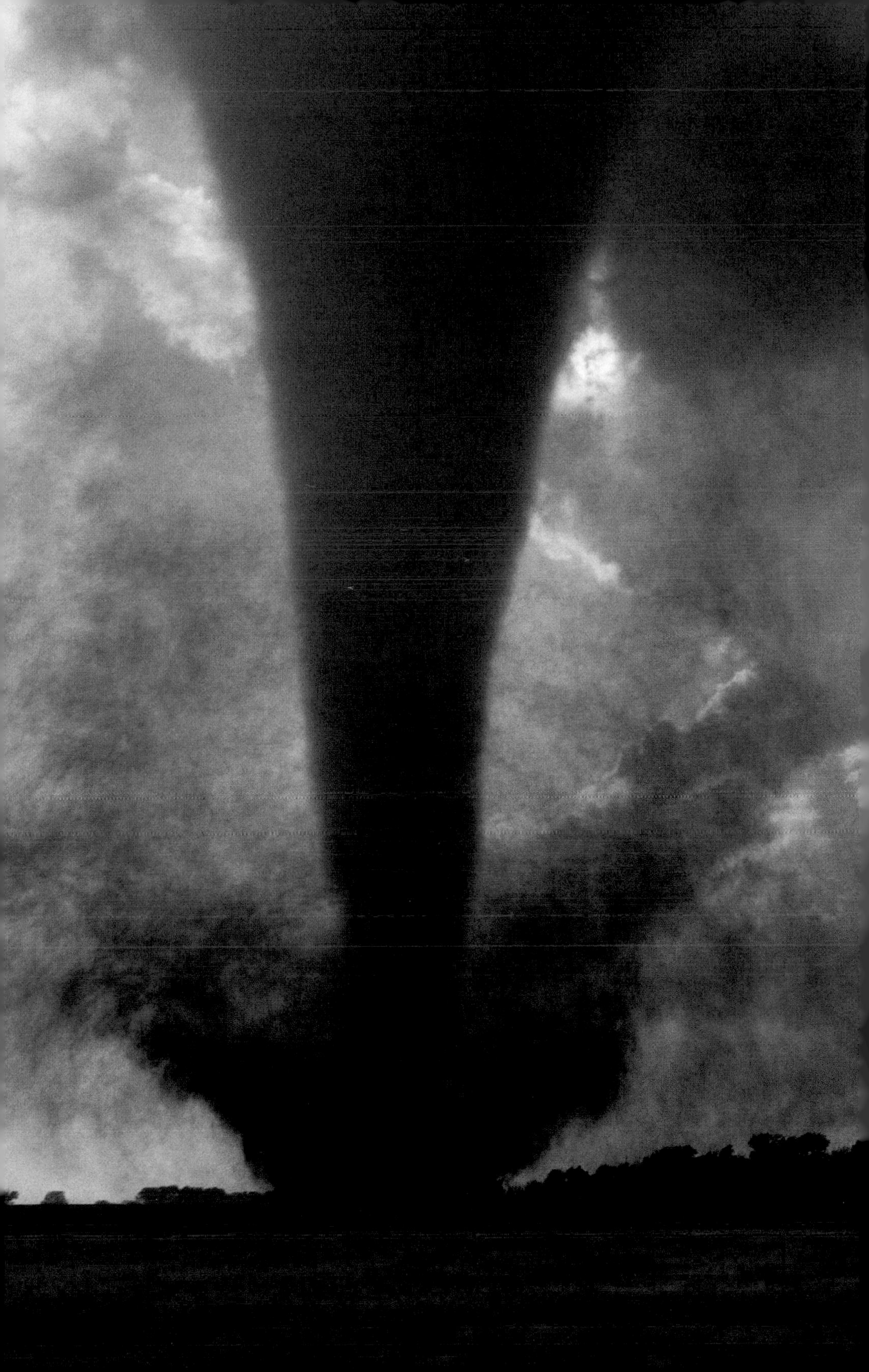

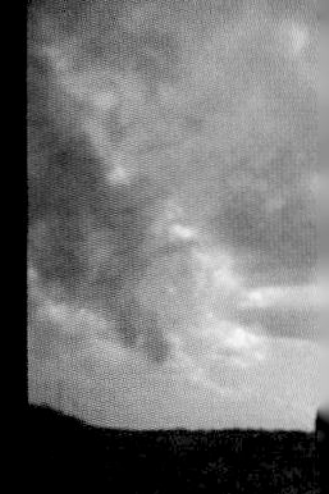

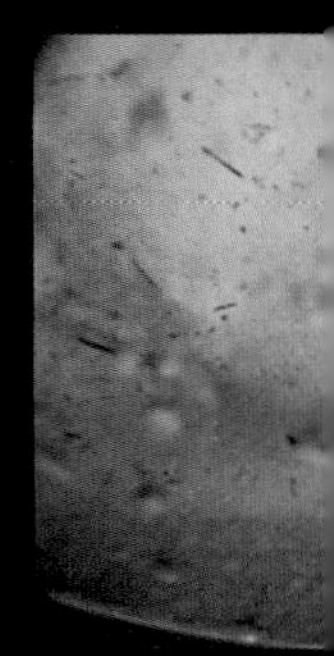

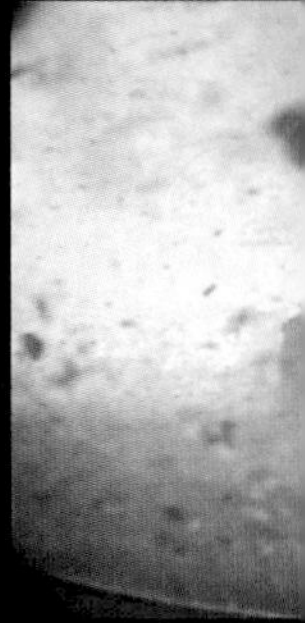

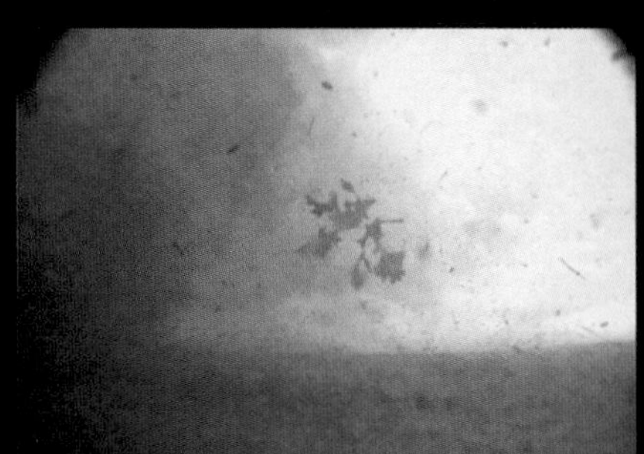
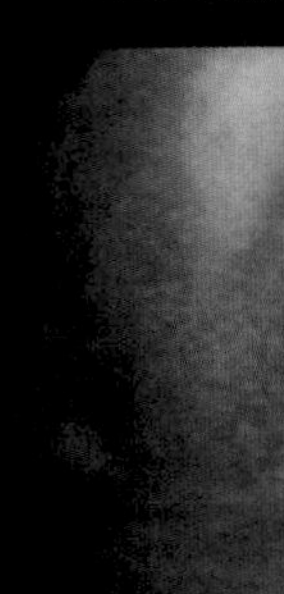

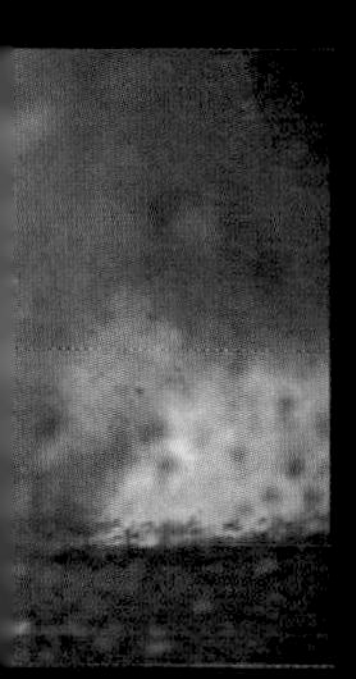

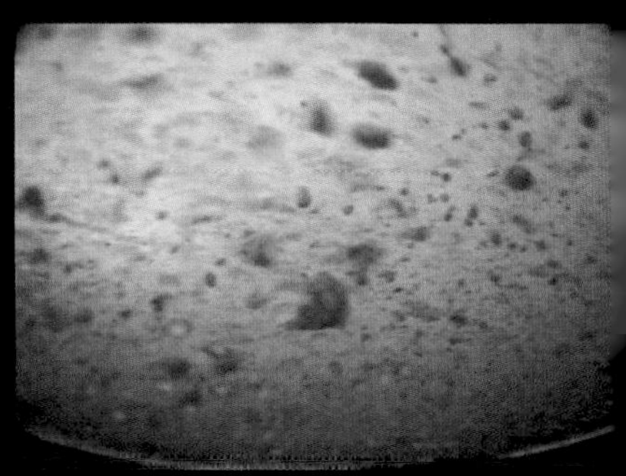

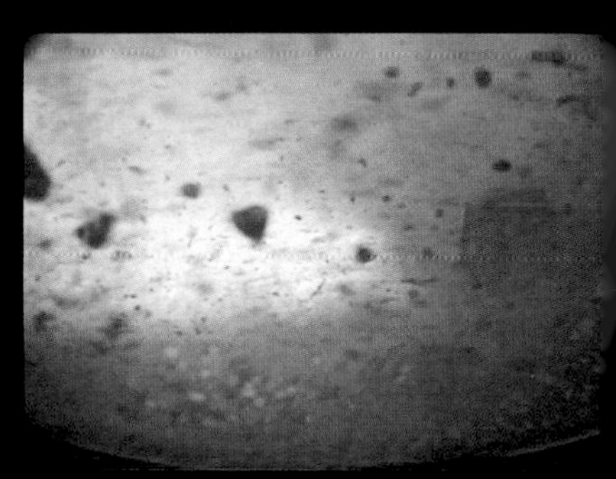

PRECEDING PAGES: *With a 360-degree view, Tim Samaras's probe documents activity inside a tornado. Each image is composed of two superimposed photographs, made milliseconds apart; the final product allows Tim to estimate the velocities of objects caught in the tornado's funnel.*

LEFT: *The sun shines from behind this tornado on May 29, 2004, the first of many produced by a prolific storm cell in south-central Kansas.*

TOP: *On June 6, 2007, storm chaser Jayson Prentice captured this panoramic view of the vast horizontal outflow of a tornado-producing storm.*

TOP: *Dark cumulus clouds and precipitation, including hail more than three inches in diameter, precede this organizing storm cell as it begins to rotate.*

BOTTOM: *On June 12, 2008, an eerie light emanates from the inside of a rotating supercell as it moves north, near Cassoday, Kansas.*

As he descended into heavier, denser air he began to feel more comfortable. He even started to believe he might survive this. But now he was enveloped by roiling clouds so thick he lost all visibility and all sensation of falling. "I felt as if I were suspended in a soft, milk-white substance and falling as though in some huge amorphous easy chair," he wrote.

He began making quick calculations, using the barely visible luminous dial of his watch. If he had ejected at about 6 p.m. at about 47,000 feet, and was falling at about 10,000 feet per minute, it should take around four minutes before he got down to the altitude where his chute would open automatically. Stay conscious, he told himself. Keep breathing.

His descent into the throat of the storm began gradually, with a mild up-and-down turbulence, like riding an elevator up a couple of floors, then dropping. He felt his body being pelted by little rocks, which he groggily realized were hailstones. Suddenly, the chute opened, and his body jerked violently as it blossomed above him.

Then, very rapidly, as he descended into the tempestuous core, the turbulence "hit me like a tidal wave of air, a massive blast . . . aimed and fired at me with the savagery of a cannon. Every bone in my body must have rattled, and I went soaring up and up as though there would be no end." Body-slammed, pounded, stretched, tumbled, and spun, Rankin—who had never had motion sickness in all his years of flying—began vomiting violently as the storm swept him sickeningly up and then slammed him violently down, over and over again. He realized that he was caught in the "heat engine" of the storm, the complex updrafts and downdrafts that might hold him aloft until he froze to death, suffocated, or was electrocuted by lightning or battered to death.

"At one point, after I had literally shot up like a bullet leaving a gun, I found myself looking down into a long, black tunnel," he wrote. He was more than likely staring down into the interior of a tornado from high aloft—a perspective rarely, if ever, glimpsed by a human. It wasn't something he could run away from. It was something he was falling *into*.

Inside an ordinary, garden-variety thundercloud, rising warm air and descending cool air form what is known as a convection cell. Convection cells are quite common; that's what happens when you boil a pan of water. Warmed water rises to the surface, cools, spreads out to the sides of the pan, then sinks to the pan's bottom, where it is warmed and rises again. This flowing circular system is a phenomenon of fluid dynamics that can occur in almost any heated liquid or gas, from a pan of chicken noodle soup to Earth's atmosphere.

Inside an ordinary "single-cell" thunderstorm (one powered by a single convection cell), the convection cell is fairly disorganized. The rising warm air collides with the sinking cool air and rain, more or less canceling each other out, which is why these kinds of storms generally peter out in an hour or less. But inside the convection cell of a supercell thunderstorm, the upwelling warm currents are organized, stable, and long-lasting because they are separated from the cool downdrafts. This is partly because powerful wind at high elevations (or wind shear) "tips" the top of the supercell forward, so that heavy rainfall does not fall directly downward, drowning out the heat engine (as it does in an ordinary thunderstorm), but instead falls *outside* the convection cell, so that the whole system can last for hours. One other big difference between a supercell and an ordinary thunderstorm: Supercells contain the powerful rotating updrafts called mesocyclones.

Everything about a supercell is, well, more super than an ordinary thunderstorm. Its convection cell can grow enormously large—up to 20 miles across—and reach much greater heights than an ordinary storm cloud. Inside the supercell, the updrafts can rise up the forward edge of the cell at 100 miles an hour or even faster, as the storm devours warm, wet air from vast distances. Downdrafts, too, are immensely powerful in a supercell thunderstorm. They are not tornadoes, but they can still blast out of the bottom, rear end of the storm at 75 miles an hour—technically, a hurricane-force wind.

The typical supercell life story goes something like this: On a particularly hot, moist afternoon, rising wet air begins to coalesce into a "multicellular" thunderstorm, a massive system containing more than one storm cell and thus more than one powerful updraft. Gradually one of these storm cells begins to move apart from the others. Because of its isolation, it begins to feed on storms from miles around, like a greedy monster. Supercells are antisocial; they do not play well with others. As the supercell greedily steals its neighbor's heat energy, it also begins exhaling downdrafts around itself, which serve to suppress the growth of nearby storms.

This monstrous inflow of moist air begins towering up into the atmosphere, to the tropopause or higher. Although the physics of vorticity is complex, the short form of the story is fairly simple: When you suck a large volume of gas or liquid through a narrow opening, it tends to spiral. That's what happens inside your straw when you drink a milk shake, and what happens when water spirals into a vortex as it escapes out the drain of a bathtub. Something similar occurs in a supercell. Huge volumes of rising air, compressed into a comparatively narrow opening, begin to rotate as they gain elevation. High-altitude wind shear (wind that tends to spiral around as it gains elevation) also contributes to this spinning motion.

GLOSSARY

Convection In meteorology, the vertical movement of heat and moisture in the atmosphere through updrafts and downdrafts, especially in an unstable atmosphere

Wind shear at the higher elevations also tends to tip the storm's top forward, so that rain falls down along the leading edge of the storm rather than straight down into its core (which would cool off and thus decelerate the heat engine). These sinking downdrafts are known as front flank downdrafts (FFDs). So at this point, the main updraft is at the rear end of the supercell.

Hail forms when supercooled water is swept aloft and then freezes; sometimes it falls, is swept up again, refreezes, and so on, growing larger with each transit. Hail size is determined by updraft strength—the stronger

A rare mother ship cloud formation moves across the Texas Panhandle.

the updraft, the larger the hailstones. They are held aloft until they either grow heavy enough to overcome the updrafts or some weakening or fluctuation in the updrafts allows them to fall. The result is a hailstorm.

Large hail is something that commonly occurs in thunderstorms, but especially in supercells. A modest, single-cell thunderstorm may produce hail up to about the size of a nickel; a multicell or cluster storm, up to about the size of golf balls; and a supercell—like everything else about supercells—produces hailstones of fabulous size, sometimes as big as softballs.

"If you spot hail larger than golf balls, you are very near a supercell's main updraft and should go quickly to a safe place," according to the National Weather Service *Basic Spotters' Field Guide,* a training booklet for storm spotters. (That is, unless you're riding in Tim's geekmobile with its rooftop hail-sampling contraption—in which case you should head straight for the center of the storm.)

Now the supercell towers up to 50,000 or 60,000 feet, an atmospheric creation of extraordinary size, grandeur, and complexity. It's a self-feeding organism, which draws in fuel from other storms at great distances; suppresses competitors; and continues to grow in power. It can last six hours or more, sometimes spawning a series of tornadoes in succession. It has been calculated that every second a mature supercell releases the energy of 20,000 tons of TNT, equivalent to the yield of a Nagasaki-size nuclear bomb.

"It may be just a little bit confusing to compare the energy released in a tornado to the energy released by a bomb," says Tim, who has spent much of his professional career studying explosions of all kinds. "The big difference is that in a bomb blast there is an explosive, near-instantaneous release of energy, whereas in a tornado the dispersion of energy is more gradual. Even so, these comparisons are accurate, and they help to demonstrate the really staggering power of a supercell thunderstorm."

It's in this mature phase of its development that the supercell may develop what's known as a mid-level mesocyclone (a powerful rotating updraft) in a region about four miles above the ground. It occurs inside the supercell's main updraft. It cycles rising air through the bottom, spinning it out at the top. As the rotating area begins to spin faster—like a figure skater drawing in her arms to increase the rate of spin—the "meso" begins to get narrower and narrower at the same time that it stretches taller and taller. The meso may begin with a diameter of seven miles and squeeze down to two or three miles, and eight miles high.

Just before the formation of a tornado, a couple of things appear to happen inside the supercell. The meso shifts toward the rear end of the supercell. The mesocyclonic rotation becomes considerably more intense; it forms what's known as a tornado vortex signature (TVS), which shows up on storm-chasing computer programs as a small funnel-shaped icon—the jackpot of storm chasing. At the same time, a tremendously powerful downdraft, the rear-flank downdraft (RFD), begins cascading down to the ground and flows outward, creating a gust front so strong it can knock over utility poles or even boxcars.

Now there is the main, rotating updraft and two primary downdrafts. The place a tornado appears most likely to form is at the intersection of the supercell's main updraft and the RFD. But precisely how the RFD contributes to tornado formation, or "tornadogenesis," remains a puzzle, which is why researchers like Bruce Lee, Cathy Finley, and others are so intent on getting better data about it.

But that's only one of many questions about tornadoes that still confound researchers.

"In fact, some of the biggest remaining questions seem to be the most basic," Tim says. "The greatest atmospheric scientists in the world can't predict exactly where a tornado will form. And we still don't know why some storms produce tornadoes and others do not. That's about as basic as you can get."

"But the question that bothers me the most is, What are tornadoes *for?*" says severe-storm meteorologist Charles Doswell. "What purpose do they serve in the atmosphere? We know the answer to this question regarding most other atmospheric processes, but when it comes to tornadoes, we don't really have a clue. In so many ways, we're still completely mystified by them."

"Tornadoes are an incredible marvel of nature and among the most fascinating scientific puzzles on this planet, one that may take the better part of the next century to unravel," adds tornado researcher Tom Grazulis.

"The first clap of thunder came as a deafening explosion," Colonel Rankin said, one so loud that had it not been for his tightly cushioned helmet, his eardrums probably would have been shattered. "The claps of thunder were not auditory sensations; they were unbearable physical experiences—every bone and muscle responded quiveringly to the crash. I didn't hear the thunder; I *felt* it."

But the lightning was even worse.

"I used to think of lightning as long, slender, jagged streaks of electricity; but no more. The real thing is different. I saw lightning all around me, over, above, everywhere, and I saw it in every shape imaginable. But when very close it appeared mainly as a huge, bluish sheet, several feet thick, sometimes sticking close to me in pairs, like the blades of a scissor, and I had the distinct feeling I was being sliced in two."

Eventually the air seemed to get smoother, and the pounding hail diminished; Colonel Rankin saw a break in the clouds below, and then, in a flash of lightning, a glimpse of green grass. When he landed in a muddy field in North Carolina, he lay there stunned and in disbelief. He was battered, bruised, freezing, bleeding, exhausted—and alive. He glanced at his watch. It was 6:40. A descent that should have taken less than ten minutes had taken forty. By that time, Lt. Herb Nolan was landing his Crusader safely in South Carolina.

Eventually Rankin cut himself loose from the chute with a big knife and stumbled to a rural highway, where he tried to flag down a motorist in the pouring rain. He needed to get medical help, immediately, but car after car slowed, paused, and then flew past without stopping. It might have had something to do with his appearance, he realized later: Looming out of the darkness at the side of the road, he was a frightful sight "in my tattered flight suit and Buck Rogers helmet, my face swollen and raw, blood caked on my face, still oozing . . . from my mouth, nose and ears, a large naked knife flashing in my hand." It took him a long time to catch a ride, but eventually a car stopped. It was a farmer, with four young boys in the backseat. The boys' eyes widened when Rankin told them he was a Marine jet pilot who had just parachuted down through the storm when his plane crashed. The boys immediately started to quarrel over who had seen Rankin first.

"Here, boys. Thanks," Rankin told them soothingly. "I owe you my life."

And then their eyes grew even wider when Rankin pulled off his battered helmet, relic of one of the most hair-raising rides in human history, and tossed it into the backseat of the car.

"There's your souvenir," he said.

9: DOUGIE DOKKEN'S BIRTHDAY CAKE

FIFTY YEARS LATER, DOUG DOKKEN WOULD CLEARLY REMEMBER THE SWELTERING North Dakota afternoon when it happened, even though he had been only seven years old at the time.

"It was like a monstrous and strange creature from another dimension that was completely oblivious to our existence," he recalled, still in awe of that long-ago experience. He and his brother Dickie, who was six, had spent almost that whole day of June 20, 1957, playing in the backyard of their home on $10^1/_2$ Street in Fargo. By midafternoon their faces were streaked with dust and sweat, because the weather was hot and sultry that day.

Others would remember that the day was also eerily still; there was barely a breath of wind to stir the damp heat that lay like a sodden blanket over the east-central Dakotas. Fargo was only 150 miles south of the Canadian border, on the banks of the Red River, which separated North Dakota from Minnesota, but summers could still get oppressively humid at that northerly latitude.

"Boys, come up here and take a look at these clouds!" their mother called, from where she stood on the front porch. It was around six o'clock, just before suppertime. The children ran around to the front of the house, scrambled up beside their mother, and then followed her gaze up into the heavens. There in the sky, to the west of the house, they could see an immense mass of roiling clouds, enormously wide, hovering like some unearthly apparition above the ground. It "appeared to be like a giant upside-down birthday cake with greenish black frosting melting upward on the sides," Doug Dokken remembered. It was rotating slowly, with striations spiraling upward into the overhanging cloud ceiling. Wrapped around the base of the "birthday cake" was a ringlike structure (which would come to be known, after this historic day, as a collar cloud).

Attached to the north end of this great cloud mass, which would come to be called a wall cloud, was a narrow horizontal cloud that appeared to be "sliding and slithering" into the birthday cake (a tail cloud). To the boys, the whole amazing formation also looked a bit like St. Nick's pipe, with a long, skinny stem attached to a bulging bowl.

"The scenes of 'mother ships' in the movie *Independence Day* and *Close Encounters of the Third Kind,* as they slowly and majestically move over the masses below, remind me of the emotion and the scene as the parent storm slowly moved toward us," Dokken later said. "People slowly and silently came out of their homes, interrupting their dinners, and gathered in stunned silence on the sidewalk and the street to see what we now call a supercell thunderstorm."

Out at Hector Airport north of Fargo, 26-year-old Ray Jensen had started his swing shift as the "warning meteorologist" at about four o'clock that day. It was his first year with the Weather Bureau. He'd grown up in the Dakotas and was familiar with this kind of moist, heavy, vaguely ominous summer weather. It "had the feel of thunderstorms, and I was almost sure there would be thunderstorms that evening," he said.

Jensen's most important job was to warn the surrounding community in the event of severe weather—thunderstorms, primarily, but also the occasional tornado (though Fargo was so far north that tornadoes were unlikely). If the weather looked like it was getting ugly, Jensen had a couple of different ways to get out the message. The weather station had a teletype connected to radio and TV stations. He could also pick up a phone linked to the stations and talk directly to the listening audience. In addition, the local weather bureau had a direct line to the regional office in Kansas City, to which he could transmit severe-weather warnings and make contact with the national Civil Defense Warning System.

It was 1957, and Dwight Eisenhower was in the White House. Nikita Khrushchev was in the Kremlin, and a couple of months later the Soviets would shock the world by successfully lofting a tiny, 184-pound satellite called Sputnik 1, about the size of a beach ball, into near Earth orbit. It was the world's first artificial satellite, and it touched off what came to be known as the space race. But down on Earth, in Fargo, North Dakota, there were few if any space age instruments available for the ordinary meteorologist simply trying to predict the weather. "Eyes, paper, and pencil" were the main tools available to local weathermen, Jensen said. Now, glancing out the airport's west window at the gathering storm, he jotted some notes about what he'd say if things started getting really bad.

GLOSSARY

Squall line A line, either continuous or with breaks, of active thunderstorms; squall lines are often poor tornado producers.

Collar cloud The circular ring of cloud that may be observed around the upper part of the wall cloud, but not the same thing as a wall cloud

Just before 6:30, Jensen was watching a dark squall line of glowering storms approaching Fargo from the west—a different part of the same massive storm system the Dokken brothers and their mother were watching from their porch—when to his surprise, a tornado funnel dropped out of the cloud and spun all the way down to the ground. He called the tower and they, too, confirmed his observation: It was a tornado, about 20 miles to the west, near the town of Mapleton. Jensen picked up the phone and read tornado warnings to the people of Fargo, then communicated the same news to the Kansas City weather station. Then he called his wife, who lived in the neighboring town of Moorhead, over the river in Minnesota, and told her to get down in the basement.

A woman in Fargo was doing the supper dishes while listening absently to the television around the same time that night. Though TV was only ten years old, and still black and white—*Leave It to Beaver* would make its debut a few months later—it had already been around long enough for people to grow jaded by it. When an announcer for WDAY-TV came on the screen to announce that a tornado warning had been issued, the woman paid it little mind. There were often a few tornado watches in the summer, but they always seemed to blow over without harm.

Just then there was a knock at the screen door and the woman's neighbor pushed her way into the kitchen with an armload of freshly picked peonies, their pendulous heads releasing a trail of pale petals onto the floor.

"They say a big storm's blowing up," the neighbor lady said. "The rain will probably ruin these flowers, and I just thought you might like a few."

The woman thanked her neighbor, found a vase, and filled it with the flowers. Then she wiped her hands on her apron and stepped out onto the front porch. The sky had grown frighteningly dark, she noticed. Cars were rushing past, faster than usual, with their headlights on.

Out at the airport, Ray Jensen watched as the squall line closed in, by now absolutely black and menacing, with an immense tornado seemingly embedded in it. The phone started ringing off the hook. He was on a separate telephone line, narrating the storm's approach to radio and TV stations as he watched the enormous cloud near, its base hovering perhaps 2,500 to 3,000 feet off the ground. Suddenly, a second tornado appeared, and it "dropped out of the cloud, a beautiful cone shape, with a sharp point on the ground," he recalled later. The dark funnel grew increasingly large, widening to about three-quarters of a mile as it moved into the Golden Ridge area, the poorest part of Fargo, on the west side of town. Then all communications went dead. Jensen could do nothing but step outside the weather station and watch as the tornado rampaged into the city, less than a mile from where he stood. Debris lofted into the black sky as it moved toward the campus of North Dakota State University.

For what seemed like ten minutes or more, Doug and Dickie Dokken stood out on the front porch with their mother, who was clutching their baby sister, watching the eerie cloud churning toward Fargo. Then, abruptly, their mother gave a little cry of alarm and herded the two boys into the house.

"I just saw a tornado come down out of that cloud!" she shouted as they ran into the house. Just then the telephone rang, and she paused briefly to answer it—it was their father, calling from work to say that he would be home as fast as he could get there, and admonishing them all to get down into the basement, *now*. The boys' mother grabbed a radio, and then they all fled down the basement stairs, followed by the whimpering family dog.

The little family cowered down there, trying to hear whatever fragmentary news came through on the radio between explosions of static. Moments later their father came home, checked in with them briefly, then went back upstairs to watch the incoming storm through the front window. A few minutes later he came pounding down the stairs.

"We've got to get into the car!" he shouted. *"We've got to get out of here! I think the tornado's going to hit the house!"*

They trooped back up the basement stairs; ran through the house, slamming all the windows shut; and ran out the back door for the car.

"I remember the winds feeling warm and then getting colder as an occasional raindrop hit me," Doug Dokken remembered. "The trees were bending to the ground in one direction, and then bending over to the ground in the other direction. It was as though the wind was coming straight down from above. The sky, instead of being greenish black, was now a white gray madly racing to the east. Bluish yellow lightning flickered horizontally back and forth in the low clouds accompanied by a continuous jet engine-like rumbling overhead, punctuated by an occasional boom of thunder.

"I remember small flocks of birds circling one way and then another as if they couldn't figure out which way to go, screen doors slamming open and shut, window shades slamming repeatedly against windows as people struggled to close them, the buzzing of air vents on the sides of attics, garbage cans blowing over and the lids coming off, rolling and tumbling along the ground."

Bernice Knudsen, a 24-year-old nursing home assistant, was driving to Fargo that evening with a girlfriend after an out-of-state visit. She lived a block from the North Dakota State University campus, but shortly before she arrived home she found herself trapped in the center of the frighteningly violent storm that Dougie Dokken's family was now fleeing.

Bernice stopped the car, and then she and her friend lay down on the floorboards as the car rocked violently in the wind, like a mouse shaking in a terrier's mouth. The storm, she remembered later, had a "mournful, whiny little voice." A huge tree limb smashed through the car's windshield. Shattered glass and rain exploded over the two huddled young women. Then they could feel the entire vehicle being lifted into the air and lofted over the roofs of houses. It landed, six blocks away, on the lawn of the NDSU field house. Incredibly, the car landed on its wheels, bounced a couple of times, and then came to a stop.

The two terrified women scrambled out of the car and ran to take cover in the field house. But the big field house doors were locked, so they clung to the door handles as the tornado raged over them, threatening to suck them into the sky.

That's when Bernice heard the whiny little voice again. "The little voice came back and rolled the car across the field house lawn and rolled it a long ways away. The wind came back and that little voice, it was trying to tear us off the door. We were straight out into the air. We were hanging on for all we could."

Finally the wind died down, and the two young women released their grip and ran for safety. Bernice realized the tornado had taken her shoes and cut the feet off her nylons without snagging a run.

When it was over, the storm had killed 13 people and damaged or destroyed more than 1,300 homes. It was later rated as an F5, the rarest and most severe kind of tornado, with wind speeds estimated at 261 to 318 mph. It is, to this day, the most catastrophic natural disaster in North Dakota history.

In the days after the storm, as people numbly picked through the rubble, they began finding scenes of unsettling strangeness. There was the incongruous serenity next to absolute destruction, like the untouched plates set out for a picnic not far from a car crushed like a soda can. There were the vinyl LP records rammed into the side of a telephone pole. The chickens almost completely stripped of feathers, terrified but still alive. Perhaps the eeriest of all was this: Inside a completely closed house, a closed desk drawer was found to be filled with wet leaves.

But the Fargo tornado was to go down in history not because of its extraordinary destructiveness or the eerie particulars of its aftermath. It was to gain its fame for two quite different reasons. One was the fact that there had been almost an hour of warning between the time a tornado was first spotted by Ray Jensen, out at the airport, around 6:30 p.m., and the time of touchdown in Fargo shortly after 7:30 p.m. Even today, in a world of Doppler radar, weather satellites, wireless Internet, and other technological wonders, the average warning time for a tornado is scarcely more than ten minutes.

Because of this extraordinarily early warning, and the majestic slowness with which the storm moved into Fargo-Moorhead (estimated at about 16 mph), amazed onlookers had time to run inside their houses to grab their cameras and photograph the oncoming storm. A few others grabbed one of the Super 8 home movie cameras that had become all the rage in the 1950s. As a result, the tornado was filmed and photographed by dozens of different people, from a variety of different angles.

It was as if, for perhaps the first time, an extremely violent tornado had paused, opened its dark shrouds, and allowed humans to catch a fleeting glimpse of its mysterious core.

The second reason the Fargo tornado went down in history was because an exceptional individual was able to see and begin to understand the secrets the storm had revealed. That man was a small, tidy, obsessively meticulous Japanese tornado researcher named Tetsuya Theodore Fujita. He was known to friends simply as Ted, but he would soon become known to the world as Mister Tornado.

A couple of weeks after the Fargo tornado, in early July, 1957, a U.S. Weather Bureau meteorologist named Ferguson Hall dropped by Ted Fujita's tiny office at the University of Chicago. Hall wanted to show Fujita some tornado photographs he'd collected on a trip out to North Dakota to inspect the damage from the storm. The destruction was phenomenal. The tornado had essentially ripped open a five-block-wide path of near-total devastation through downtown Fargo and across the river to the neighboring city of Moorhead, Minnesota.

Fujita, then 38, had a temporary position as a visiting researcher at the university, working under Horace Byers, who had gained national renown for his work with the Thunderstorm Project (see Chapter 8). Though Fujita was still a junior associate on a meager salary—he still couldn't afford to bring his family over from Japan—he was quickly developing a reputation as a brilliant meteorological investigator.

To Fujita's great excitement, it appeared that so many photographs had been taken, by so many different people, from so many angles, that it might be possible to put together a sequence of images that would show key aspects of the architecture of the storm, as well as how it grew, matured, and dissipated. This wealth of visual data might allow him to "see inside" the whirlwind and perhaps get a glimpse of how it worked. He was particularly intrigued by the photographs Hall showed that seemed to depict an immense, ominous-looking cloud mass from which the tornado appeared to have descended—Dougie Dokken's "upside-down birthday cake."

Fujita's elation stemmed from the fact that, for a variety of reasons, tornadoes are extremely difficult to study. For one thing, they're rare. Only about one in a thousand thunderstorms produces a tornado. Fewer than one percent of Americans can expect to encounter even the weakest tornado in their lifetime. And "even in the most tornado-prone areas of the country, a home can expect to be hit only about once in a thousand years," according to tornado researcher Tom Grazulis.

"When you're out there actually trying to *find* tornadoes like we do, using all the most advanced technology there is, plus the best judgment of experienced severe-storm meteorologists like Carl Young—and you *still* can't catch the crazy things, you get a sense of how rare tornadoes actually are," says Tim. "We've gone months at a stretch, even whole seasons, without really getting close to one."

For another, tornadoes are often almost as ephemeral as a summer thundershower. Although they are capable of lifting a tractor trailer truck off the ground, they are, after all, merely a momentary disturbance of wind, dust, and water vapor. Unlike hurricanes, which can last for days or even weeks, the average tornado lasts only about 20 minutes. Most last less than ten. But even if you get close enough to see and hear a tornado, chances are that it's wrapped in a thunderstorm, one of the most blinding and dangerous environments on Earth, loaded with high wind, battering hail, and 30-million-volt lightning.

To get close enough to actually make contact with the tornado, to get your hands right down inside it like a medical student might plunge his hands into a cadaver, is nearly impossible. And it was even more nearly impossible back in the late 1950s. Even today, getting direct measurements like wind speeds or barometric pressure is extremely difficult.

Now Fujita seemed to have an unparalleled opportunity to actually *study* a tornado, almost as if he'd been able to drag it into his lab. Byers agreed to let Fujita undertake a complete "mesoanalysis" of both the tornado funnel and the surrounding clouds by visiting the devastated area and collecting all the available photographs and film he could find. (This is known as a photogrammatic study.) Fujita visited Fargo three times, and with the help of WDAY-TV weatherman Dewey Bergquist collected a total of over 200 photographs, color and black-and-white, taken from about 50 different locations, as well as several short Super 8 movies.

Then Fujita began the painstaking process of reducing each photographic image to the same dimensions; establishing precisely where and when each photograph was taken; and determining the angle from which the photo had been shot. He also meticulously, almost obsessively, walked the damage paths on the ground in Fargo.

This was tedious work, almost completely unaided by modern technology (except for photography and film). To make sense of the chaotic and widely scattered evidence after a tornado touchdown, meteorologists of that day had to apply sheer brainpower, particularly the power of deductive reasoning. Fujita excelled at all these tasks—not unlike that other brilliantly deductive detective, Sherlock Holmes.

"My mind rebels at stagnation," Holmes declares, in a famous novel called *The Sign of the Four.* Sitting in his musty bachelor lodgings on Baker Street, Holmes exhales a ring of blue smoke and leans over the table toward his compatriot Dr. Watson, his fingertips touching.

"Give me problems, give me work, give me the most abstruse cryptogram or the most intricate analysis, and I am in my own proper atmosphere. . . . I abhor the dull routine of existence. I crave for mental exaltation!"

The ideal detective, Holmes goes on, possesses three qualities: knowledge, observation, and deduction. Holmes's knowledge of a variety of things—from the ink-absorption rates of various kinds of writing paper to the symptoms of poisoning by exotic adder bites—is staggering. His powers of observation are legendary. And his capacity for deductive reasoning is breathtaking. By observing a fleck of reddish soil on Watson's shoe, and salting in a couple of other clues, Holmes is able to deduce that Watson has just dispatched a telegram from the Wigmore Street Post Office. When Watson gasps in disbelief, Holmes uncorks one of his most famous remarks:

"How often have I said to you that when you have eliminated the impossible, whatever remains, however improbable, must be the truth?"

It was a comment that might as well have described the relentless logic that Ted Fujita applied to the problem of tornadoes. Fujita has been called by *Weatherwise* magazine "probably the best meteorological detective who ever lived," a man who, like the fictive Holmes, consistently uncovered the clues that others had overlooked.

"Ted had the ability to take limited information and make plausible deductions as well as anybody in our history," says Charles Doswell, a former research meteorologist at the National Severe Storms Laboratory (NSSL) in Norman, Oklahoma. "It's the hallmark of genius. It's almost impossible to overestimate his impact on the scientific understanding of tornadoes."

"There was an insight Fujita had, this gut feeling," adds Jim Wilson, a senior scientist at the National Center for Atmospheric Research in Boulder, Colorado. "He often had ideas way before the rest of us could even imagine them."

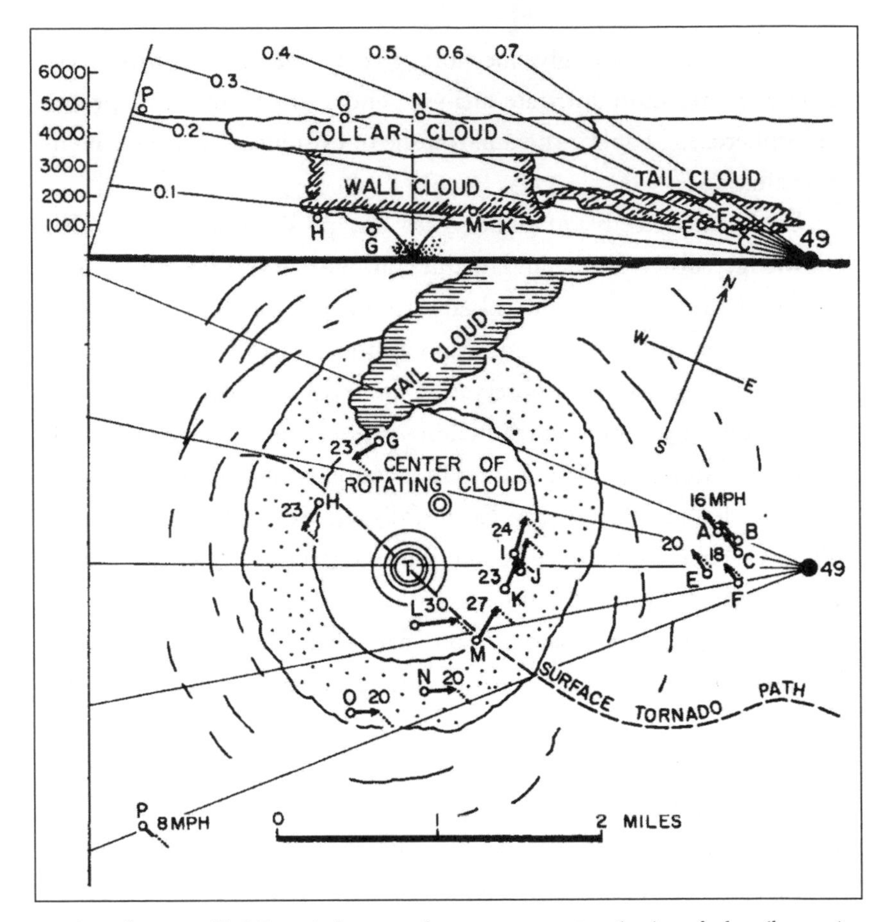

Tornado expert Ted Fujita's drawing shows a cross section (top) and plan (bottom) view of the 1957 Fargo tornado and depicts the height of the collar, wall, and tail clouds, as well as the tornado's direction and speed.

Fujita possessed another special skill that was to prove critical to his later success: the ability to brilliantly visualize things.

"The core of science is to 'see,' to 'visualize' the physical concepts that weave the observational evidence together into a coherent explanation of the phenomena responsible for those observations," explains Leslie Lemon, a former researcher-meteorologist at the NSSL. "He had a rare gift to see what most of us could not see."

Which brings to mind another famous quote from Sherlock Holmes, when he chastised Watson for his careless attention to detail: "You see, but you do not observe."

Three years after the North Dakota storm, Fujita published a paper called "A Detailed Analysis of the Fargo Tornadoes of June 20, 1957." This study, with its intellectual clarity, its Holmesian logic, and its elegant charts and drawings, was "a landmark in storm analysis that in some ways is unequaled even today," Tom Grazulis has written.

What Fujita quickly realized when he began piecing together the story told by the photographs and film was that "some of the pictures were taken long before the tornado funnel started dropping from the base of a huge rotating cloud. After hearing radio and television reports of the U.S. Weather Bureau's tornado warning, some people apparently mistook the black rotating cloud, at least ten times larger than a tornado in horizontal dimensions, for the tornado itself and began taking pictures of the cloud."

The second thing that quickly became obvious was that the Fargo storm was not just one tornado. It wasn't even three, as was originally thought. It was *five* tornadoes in succession, each one leaving a continuous damage path up to 11 miles long and up to 700 feet wide.

"Further study also confirmed that these tornadoes were produced by a rotating cloud something like a miniature hurricane," Fujita wrote. Though he had not used the term "supercell" in this paper—it had not yet been coined—he had just described it. (In fact, in a 2001 scientific paper about Fujita's contributions, meteorologists Gregory Forbes and Howard Bluestein wrote, "It is noteworthy that in recent years, close-range mobile Doppler radar-reflectivity images of tornadoes look like those of scaled-down hurricanes.")

The fact that this "miniature hurricane" had seemingly given birth to what he called a "family" of tornadoes, one after another—this was a startling new observation. It wasn't that people had not seen, or even photographed, tornadoes in which there appeared to be multiple vortices before, Charles Doswell says. It was that Ted Fujita "was able to recognize the importance of things that other people had seen but overlooked. He was beginning to recognize that some thunderstorms are special, they are different—they're supercells."

He had not just seen, he had *observed.*

Fujita began his paper by laying out the general weather situation around Fargo on the afternoon of June 20. He constructed weather charts at three-hour intervals, based on information from local weather stations using techniques developed in the Severe Local Storms Project at the University of Chicago. At the time the storm hit the city, "a small mesoscale thunderstorm high accompanied by heavy convective activity was approaching Fargo from the northwest."

As it moved toward Fargo, the thunderstorm—actually, a supercell—towered so high that a radar station 205 miles away was able to detect cloud tops up to 75,000 feet high.

The first tornado was spotted about 4:30 that afternoon, west of Fargo in the farmland near Wheatland. Several farmers saw a small, dust-filled whirlwind, perhaps a hundred feet across, which "appeared like a rope or a light-colored snake" twisting into the sky. This first tornado was so weak and ephemeral there was some question whether it was a true tornado at all. The second tornado appeared at almost the same moment the first one vanished. It was much darker than the first, with an east-moving, cone-shaped funnel. Based on the photographs and a short home movie, Fujita estimated the funnel's diameter at about 400 feet.

A woman who looked up at it as it passed over her farm in Casselton, North Dakota, made one of those surreal observations for which tornadoes are famous. "When she looked up she saw a black bag hanging down from a dark cloud," Fujita reported. "At its center was a hole, inside which circled a number of objects resembling tree branches. In spite of such a spectacular display aloft, nothing particular was felt on the ground."

It was the third tornado, the one that came to be known as the Fargo tornado, that was the real monster. Because so many people in the towns of Fargo and Moorhead photographed the rotating cloud as it approached from the west, the tornado's entire "life history" was documented, from formation, to maturation, to dissipation. This, also, was new—not the fact that tornadoes have life cycles, but the fact that, for the first time, this life cycle had been documented with photographs.

In one particularly harrowing photo sequence, the funnel can be seen descending from the cloud base all the way down to the ground in about 14 seconds. After the touchdown, the diameter of the funnel increases rapidly, then it lifts off the ground, as if the funnel had been sheared off at the bottom.

A fourth tornado developed near the town of Glyndon, severely damaging trees along the Buffalo River and demolishing a farm. Farmers saw the funnel gradually shape-shift into a light-colored rope as it moved northeast and then vaporized into the air, leaving a damage path ten miles long. The last tornado was a dark cone, which moved across the countryside on a northeasterly path, destroying the Gol farm and stopping an electric clock at 8:05 p.m. when it shattered a high-voltage pole. Observers said the tornado gradually morphed into a big, black hose, and finally into a long, twisting rope that disappeared over Stinking Lake.

In his studious, meticulous way, Fujita mapped out the precise path of each of the five tornadoes through the five different towns, including sketches of the shape of the funnel as described by observers. Then he mapped out the damage paths of the five tornadoes. The damage paths of the Fargo and Glyndon tornadoes had a narrow tip widening into a long, fat body that narrowed again into a long, skinny tail. Fujita's diagram looked a bit like a fat leech.

Taken together, their damage paths extended about 70 miles. Though the Fargo tornado lasted only about 20 minutes, the entire tornado family lasted about three hours and 45 minutes.

One of the most enduring legacies of Fujita's paper was that it described and gave names to various cloud features of a supercell that still bear these names today. Little Dougie Dokken's birthday cake—the immense cloud mass hanging down below the cloud base—Fujita called a wall cloud. He gave the term "collar cloud" to the ring-shaped formation that encircled the rotating wall cloud at the top, where it was suspended from the cloud base. (If the wall cloud were an upside-down birthday cake, the collar cloud would be the rim of the plate upon which the cake was sitting.) The structure that Doug Dokken thought looked like the stem of St. Nick's pipe—a narrow, horizontal formation streaming out to one side of the storm—Fujita called a tail cloud.

Fujita intensely analyzed one short, 810-frame Super 8 film the way the FBI analyzed the Zapruder film of the Kennedy assassination. By tracking the vertical progress of a certain irregular feature on the side of the wall cloud, Fujita deduced that there was an extremely rapid vertical velocity—from about 80 feet per second (55 miles an hour) at the 3,000-foot level to as much as 200 feet per second (135 mph), and perhaps much higher, at 10,000 feet. This rate of acceleration was far higher than anyone had previously realized. It was also much faster than the rate of spin in the surrounding cloud mass, which was revolving at a relatively leisurely rate of about 25 miles an hour.

Fujita also deduced that the ring-shaped collar cloud was rotating at about 10 to 25 miles an hour, and that the top of the ring was rotating faster than the base of the ring.

Using all of this photographic and film imagery, Fujita created a series of drawings, which he then turned into a short animated film. Projected on a screen, the film clearly showed the features of the rotating wall cloud and its successuve stages as it spun off tornadoes, one after another, across the North Dakota flatlands and on into Minnesota. It was the first time in history that the life story of an entire family of tornadoes could be seen, in detail and in slow motion.

In some ways Fujita's little tornado cartoon was like an old Charlie Chaplin silent film—flickering, crude, and silent, but filled with the promise of much greater things to come.

10: THE MYSTERY OF FLIGHT 66

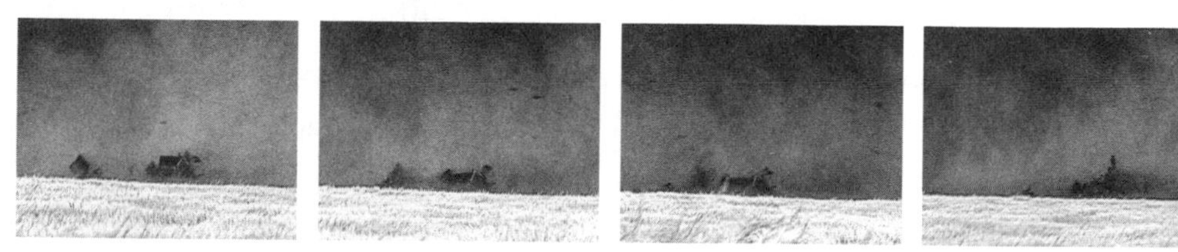

IN AUGUST 1953, A TINY, TIDY JAPANESE MAN WITH ROUND EYEGLASSES, A NEAT suit, and a beatific smile boarded an airplane in Tokyo on a one-way trip to the United States. The ticket had cost Tetsuya Fujita $650, the equivalent of 13 months of his salary at the Kyushu Institute of Technology, in Japan, where he was then working. Although he was 33 years old, Fujita was anxious; this was to be the first airplane ride of his life. In fact, he'd never been outside Japan before.

Nevertheless, Fujita was excited beyond measure. The distinguished American meteorologist Horace Byers, famous for the Thunderstorm Project (see Chapter 8), had invited him to come to America for a two-year research post at the University of Chicago. Fujita had recently completed the Japanese equivalent of a doctorate, the Sc.D., with a thesis that brilliantly analyzed the structure of hurricanes. Severe weather and all its mysteries had become his calling.

On the airplane across the Pacific, Fujita eased his anxieties by sketching the clouds and plotting the time line of their transformations, including a magnificent cumulonimbus storm tower that bumped and rattled the airplane as they passed through the turbulent air. When he arrived in San Francisco and boarded the train for Chicago, Fujita was carrying $22, all that was allowed under postwar currency restrictions. For the three-day trip, he bought a package of Fig Newtons and some Coca-Cola. Accustomed to making a little bit of food last a long time, he rationed the fig bars to last until he made it to the Midwest, where he was met at the train station by Byers's secretary. Then he went out and celebrated with a meal that cost $1.50.

Tetsuya Fujita's arrival in America was to become a historic moment in the history of tornado science. But the story had begun many years earlier, during his boyhood growing up on the Japanese island of Kyushu, a place he described in a self-published memoir as "my wonder island," with two active volcanoes, deep forested mountainsides, and tidal estuaries rich with sea life.

From a very young age, Fujita had a mind that was a bright star—inventive, inquisitive, and unfettered. In the tradition-bound culture of Japan, which prized obedience above almost all else, he was a kind of intellectual rabble-rouser. When he was 16, he went with a school group to visit Yabakei Rapid, a river gorge where a Buddhist monk named Zenkai had dug the Aono Domon, or Blue Tunnel, through a cliff overlooking the river. The tour guide explained that it had taken the monk 30 years to accomplish this feat, using only a hammer and chisel.

"Isn't that admirable?" Fujita's teacher asked.

"No," the boy piped up. "If I had been asked to dig a tunnel, I would have spent the first 15 years inventing a digging machine. Then, when the 30 years was up, I could leave behind a tunnel *and* a new digging machine."

Because of his lack of appreciation for the monk's spiritual accomplishment, young Fujita failed the class.

A few years later, needing to find an easy way to multiply in algebra class, he painstakingly "invented" a cardboard slide rule that included fractional numbers. When he proudly showed the slide rule to his algebra teacher, the teacher, unimpressed, told him, "You spent over ten

Tetsuya "Ted" Fujita, shown here with his tornado simulator at the University of Chicago, was a pioneer in the field of meteorology whose careful analysis of severe weather, especially tornadoes, led to his sobriquet: Mr. Tornado.

days to generate the logarithms that you will be learning in less than ten minutes in your regular class next month."

Undeterred in his self-directed explorations, Fujita became interested in topographical maps and began mapping everything from his schoolyard to the surrounding mountainsides of Kyushu. He built a one-dollar telescope out of a cardboard tube and an old long-focus lens, and began mapping the rotation rate of sunspots on the surface of the sun. He determined that the rotation period was 25 days at the Equator and 27 to 29 days at the higher latitudes, which led him to the profound question, "Why does the sun rotate faster at the Equator?"

He also began exploring the earth below, and one day he and a friend stumbled upon a previously undiscovered limestone cave in the woods. They came back with flashlights and survey chains, and Fujita drew a precise and beautiful map of the cave, which was 160 feet deep and contained lovely, multicolored, curtain-shaped stalagmites that looked to him like the aurora borealis.

Eventually Fujita's scientific initiative was noticed, and when he graduated from Kokura Middle School in 1939, he received the school's Rika-Sho (science award)—the first honor of his life. Unfortunately, Fujita's father had died two months earlier, so he wasn't able to witness his son's success. Nevertheless he was present, and in the most fateful way.

"Tetsuya," Fujita's father had said before his death, "I want you to enter Meiji College even if you are admitted to the Hiroshima college for teachers."

As fate would have it, Fujita was accepted at both schools but followed his father's wishes and enrolled at Meiji College. Had he attended the college in Hiroshima, very likely he would have been incinerated when the first atomic bomb exploded over the city six years later, on August 6, 1945.

At Meiji College Fujita enrolled as a student of mechanical engineering, completing a thesis on the subject of impact forces. For it he measured the impacts of various sizes of steel balls colliding with different materials at high speed, amplified and recorded by a high-speed optical oscillograph. But his love of mapping continued, and for one professor he drew beautiful bird's-eye views of four volcanic calderas, based only on topographical maps. After a couple of months, "my eyes began seeing contour maps as if they were three-dimensional mountains," which was excellent training for his future work in visualizing 3-D pictures of tornadoes, based on scanty or fragmentary evidence.

Fujita graduated from Meiji College in 1943, six months early—and in the middle of the Second World War, when the Japanese Army was steadily retreating from the Pacific front. The day after graduation, he accepted a teaching job in the school's physics department, at the glorious salary of $17.64 a month.

In July 1944, as the U.S. Marines invaded the islands of Saipan, Guam, and Tinian, Fujita received a small contract from the Japanese Navy to figure out the positions of U.S. aircraft by using three-dimensional triangulations. Because he had to include Earth's curvature and various weather conditions in his calculations, he started becoming very interested in the meteorological aspects of the global atmosphere.

But there was no way to avoid the ever encroaching war. In March 1945, Fujita happened to be visiting Tokyo and staying at a friend's house when screaming waves of B-29 bombers appeared in the sky, firebombing the city with incendiary devices that torched 230,000 houses overnight. In his memoirs, he recalls hearing the air-raid sirens, seeing smoke-filled reddish skies, and finding unexploded incendiary devices buried in the gravel road near his friend's house the next morning. But he makes no mention of the psychic trauma of this experience. He records only that, on the train from Tokyo back to his wonder island of Kyushu, "I began to think about a mechanical/electrical analog computer operated by converting mechanical quantities into electrical signals." In a sense, physics and engineering had become his new wonder island, his refuge from the fear and horror of war.

On August 6, 1945, an American B-29 bomber called the *Enola Gay* dropped its atomic bomb over the city of Hiroshima, leveling more than five square miles almost instantly. Three days later, on August 9, a second atomic bomb was dropped over the city of Nagasaki. Once again, Fujita narrowly escaped death. The original target for the Nagasaki bomb was Kokura Arsenal, less than three miles from his college, where he and his students were cowering in a bomb shelter not far from the physics building. They could hear the air-raid sirens going off all around them as an American B-29 bomber called *Bockscar,* carrying a 10,000-pound atomic bomb—heavier than the one dropped on Hiroshima—crisscrossed the skies overhead, seeking an opening in the heavy cloud cover that blanketed the city. Unable to find one, the bomber was diverted to Nagasaki. Fujita and his students had been saved by the atmosphere.

A few weeks later, after Emperor Hirohito and the Japanese government surrendered on September 2, Fujita joined a group of faculty and students from the college who were dispatched on a "ground-truth mission" to what remained of Hiroshima and Nagasaki. Again Fujita fails to record anything about the hellish scenes he witnessed in these two ruined cities, where an estimated 120,000 people were killed instantly by the fireball and blast wave. "I visited both Nagasaki and Hiroshima, noticing that the train fares to these two cities were identical," he writes, in his serene, eccentric, and laser-focused way. As he trained his attention on the mechanics of the bombing, he was most captivated by a physics puzzle: At what altitude was the bomb detonated? Using his expertise in impact forces, and the triangulation techniques he'd developed working for the navy, he tried to reconstruct or envision beams of radiation by studying burn marks on objects on the ground. But the shock wave that followed the fireball had dislocated or destroyed many of these objects.

Finally he discovered a crescent-shaped burn mark on the inside rim of a cemetery flower pot far enough from the fireball to be undisturbed. By collecting and carefully recording many of these burn marks from flower pots in many different cemeteries, he was able to reconstruct the angle of radiation and deduce that the fireball was detonated about 1,560 feet above the ground at Nagasaki, and 1,590 feet at Hiroshima. Because the elevations were so nearly identical, he deduced that the U.S. Air Force must have had good barometric pressure data for Japan, even though there was a complete blackout of weather reports throughout the country during the war. Later he learned that a U.S. observation aircraft had departed from Tinian an hour in advance of the bomber and used a dropsonde (a parachute-guided weather sensor) to measure pressure and temperature for each bombing.

Fujita was also fascinated by the gigantic starbust pattern of each of the two blast zones, centered on what appeared to be the bomber's ground-zero target for each site: a bridge at Hiroshima and a cathedral at Nagasaki. In each case, the blast waves flattened nearly everything in an outburst pattern—trees, houses, bamboo, telephone poles. It was as if the blast pattern were a graphic representation of the impact forces unleashed during the explosion; if the debris could be "read" like a book, it might reveal things about the explosion that were not obvious. Although he noticed these patterns in the apocalyptic blast zones in postwar Japan, it was not until years later, in America, that he would learn to read patterns of destruction after thunderstorms and tornadoes better than anyone had ever done before.

Although the Japanese Army had not surrendered in 2,000 years—there was no word for "surrender" in the Japanese language—after the Nagasaki explosion, imperial Japan surrendered unconditionally to the Allied forces. The country lay in ruins, including the postwar economy. Because his salary could barely pay the bills, young Professor Fujita and the rest of the faculty were given "rest days" as compensation. Fujita would receive his miserable pay in cash, twice a month, in a brown, sealed envelope.

Eventually Fujita decided to leave his prestigious but poorly paid job as a college physics teacher, and he won a two-year grant to educate grade school teachers. He decided that weather science was a choice topic, partly because it was cheap (it took only pencil and paper), partly because lots of teachers were interested in the subject, and partly because he himself had grown increasingly fascinated by the mysteries of the global atmosphere.

It was as part of this work that Fujita had a fateful encounter with a ferocious thunderstorm during the summer of 1947. He and a colleague from the Fukuoka weather service had gone up to Seburi-yama, where Japan had a small weather station and the U.S. Air Force was installing a radar tower. They backpacked up the steep trail just as billowing cumulonimbus cloud towers began massing in the sky. Shortly after they reached the little substation on the summit, dark clouds moved in, with rumbling thunder, high wind, and flashes of cloud-to-mountain lightning. As the tiny building shook in the wind, and rain seeped in through the leaky roof, Fujita took meticulous records of wind speed, temperature, and air pressure.

Afterward, he decided to do a complete "microanalysis" of this particular storm and began visiting all the other weather stations on and near Seburi-yama to collect the data they had. There were no copying machines in Japan in the 1940s, and no national archive of weather data, so Fujita had no choice but to visit each station and make copies of the data, by hand, on tracing paper. Next he arranged each of these data pictures, with their great swirls and rolls of air pressure and temperature gradients, into ten-minute intervals, arranged on the page in a succession of small panels, almost like a cartoon. Whether by the hand of nature or the hand of Fujita, each of the panels was lovely, akin to a Japanese brush painting of a wave caught in midair. It was an exquisitely rendered cartoon weather story of a thunderstorm.

By arranging the data this way, it became apparent that a thunderstorm was a miniature dome of high pressure. The rise and fall of surface pressure as it passed over a fixed location was nose shaped, so Fujita titled his graphical cartoon "Nose of a Thunderstorm." He also deduced that there was a significant downward current of cooler air, or a downdraft, during the height of the storm. (The word *downdraft* was not used at the time, Fujita pointed out in his memoirs, because "nobody in Japan in 1948 thought about a downward current in a thunderstorm.") Fujita also deduced that air was being sucked up the leading edge of the storm and discharged as a downdraft out the rear. He had sketched out the structure of a thunderstorm, with a general explanation of how it worked. He summarized his findings in a paper called "Raiu-no-hana" ("Thunder-nose") and presented it to a meteorological society in Japan. It was received with thunderous . . . silence.

It was at about this time that Fujita got his first chance to survey the damage after a tornado, known to the Japanese as *tatsu maki,* which means "dragon swirl." He and his fiancée, Tatsuko, went to see the damage in the Enoura district and found a six-mile-long path of destruction. The twister had originated as a waterspout in Ariake Bay, plowed several miles through rice fields, and then disappeared into a bamboo forest. When he analyzed the weather data, Fujita was intrigued to discover that as the storm passed over the Saga weather station, first there was a significant drop in barometric pressure, which was followed by heavy rainfall. "Evidently," he mused, "the tornado was associated with a prefrontal shower."

Once again, Fujita began hand-gathering and hand-drawing data about this particular severe weather event—this time, with the help of his new bride, Tatsuko. A year later, he completed a 41-page report called "Micro-analytical Study of Cold Front."

He hadn't given up on the thunder-nose and downdraft concepts, though, and in 1949 he gave a talk about these ideas to the Fukuoka weather service. Afterward, an employee of the weather service told him that he'd found an interesting, related paper in a trash can at the U.S. Air Force installation on Seburi-yama. The paper was called "Nonfrontal Thunderstorms," and its author was Horace Byers of the Institute of Meteorology at the University of Chicago, in the United States.

It was one of those fateful coincidences, of which Fujita's life was so full: The weather service employee handed him the paper he'd fished out of the trash, and when he read it, Fujita felt he had found a kindred soul. He also felt that it was necessary to translate his own papers and send them to Byers. He managed to locate and purchase a used typewriter for 20,000 yen ($55.56), which was 2.7 times his monthly take-home pay. Somehow the young professor and his wife survived for a couple of months while, with one finger, Fujita tapped out halting English translations of both his cold-front paper and the thunder-nose paper. In August 1950 he mailed the cold-front paper to Byers, thinking it would be the more interesting of the two.

Three months later, Byers responded with a warm letter, concluding: "You are to be congratulated on a very careful analysis of the small-scale features in a cold front. This problem is attracting a great deal of attention in the United States at the present time, and the U.S. Weather Bureau has a special project to investigate these smaller disturbances. It is known that they lead to the formation of tornadoes which, as you know, are very common in the west-central and southern United States."

Encouraged, Fujita sent along the thunder-nose paper as well. A few months later he received a reply from Byers that took his breath away. In fact, he would later include this letter at the beginning of his memoirs, calling it "the most important letter I received in my life," because it was to lead to a faculty appointment at the University of Chicago and the beginning of his work as the premier tornado scientist in the world.

"I have looked over your paper, 'Micro-analytical Study of Thunder-Nose,'" Byers wrote, "and note that in view of the fact that you were not familiar with the work of the U.S. Thunderstorm Project on this subject your conclusions are highly valuable and really represent an independent discovery of some of the factors derived from our work. In particular you deserve credit for noting the importance of the thunderstorm downdraft and outflowing cold air."

GLOSSARY

Downdraft A small column of air that descends rapidly to the ground and creates downbursts

Climatology The science that deals with climate and climatic conditions

By dint of meticulous observation, exhaustive data collection, and implacable deductive logic—and on a pauper's salary—Fujita had made the same discovery as the lavishly funded and technologically advanced Americans, who had made use not only of research aircraft but also a new supersecret technology developed during the war: radar. Within months, Fujita was on a plane heading toward a fateful position in the Department of Geophysical Sciences at the University of Chicago.

During his first two years in Chicago, while Tatsuko stayed home with their infant son, Kazuya, Fujita began working with meteorological researcher Morris Tepper at the U.S. Weather Bureau in Washington, D.C. The focus of their research was severe weather. Fujita had become particularly fascinated by the power and behavior of that most American of phenomena, the tornado. As he recalled later in his memoirs, as a child growing up in Japan he had been taught that the "four fearful things" were, in order of fear: *zishin* (earthquake), *kaminari* (lightning), *kaji* (fire), and *oyaji* (father). But in America, he wrote, the four fearful things might more appropriately be tornadoes, lightning, fire, and crime.

He was intrigued by the fact that almost three-quarters of all tornadoes on the planet occurred in the United States. Furthermore, 90 percent of the really severe tornadoes (those that would later be categorized as F4 or F5 on the Fujita scale) occurred in the United States, most of them on the Great Plains.

Why? What conditions led to their formation? Why were they so difficult to predict? How did they work? What caused them to disintegrate and die?

While measuring temperature and pressure changes in storm systems in Oklahoma and Kansas, Fujita and Tepper began referring to the relatively small-scale, counterclockwise rotation beneath powerful storm clouds as mesocyclones. They believed that the mesocyclone had something important to do with the formation of tornadoes, which gave forecasters an important new clue about how to predict a tornado. They also began creating a new weather vocabulary to describe their findings. The word *meso* in mesocyclone meant "medium-size," or any weather system up to 250 miles wide. The same with mesolows (medium-size low-pressure areas) and mesohighs (medium-size high-pressure areas), which Fujita now carefully began mapping. The reason this was significant was that America's great prairie storms did not seem to arise from large-scale or "synoptic" weather systems, but rather from these midsize phenomena. By limiting their focus to the correct scale, Fujita and Tepper felt they might at last begin to unravel some of the mysteries that surrounded tornadoes.

In 1956, Fujita returned to Japan to fetch his wife and son and came back to Chicago. Now he had a new title: visiting research associate at the University of Chicago. And he was embarking upon what would become a 56-year career, in which he would advance the human understanding of tornadoes more than any other single individual.

In a 2001 scientific paper outlining Fujita's achievements, meteorologists Gregory Forbes (one of Fujita's former students, who is now the severe-weather expert at the Weather Channel) and Howard Bluestein (a professor of meteorology at the University of Oklahoma) point out that many of these achievements involved photogrammetry, the science of making maps and calculations from photographs—the discipline he was perfecting during his analysis of the 1957 Fargo tornadoes. In an age when meteorologists are more likely to run a mathematical model on a supercomputer than painstakingly draw a diagram by hand, this is remarkably low tech. But what he deduced is even more remarkable.

One of Fujita's core contributions was to undertake compiling a massive tornado climatology database—the "big picture" of where tornadoes occur all over the world. Out of this database, he created a

U.S. tornado climatology map for the 70-year period from 1916 to 1985, a database that is still widely used today. It shows, with stunning simplicity, that the United States is the undisputed tornado capital of the world—there are three times more tornadoes in the U.S. than anywhere else on the planet.

In the course of doing this work, Fujita quickly realized that most of the earlier tornado studies were simple tabulations of how many tornadoes had occurred in such and such a state or area. A minor dust-up was registered the same as a tornado that caused savage devastation.

To remedy this shortcoming, Fujita set out to devise a scale that would rank tornadoes by level of damage and wind speed. In a paper published in 1971, he and Allen Pearson, head of the National Severe Storms Forecast Center (now the Storm Prediction Center), introduced findings that would lead to what would be known as the Fujita scale, the F-scale, or sometimes the Fujita-Pearson scale. The scale was divided into six categories of increasing severity. Although each damage level is associated with a wind speed, it's really a scale that measures *damage,* since calculating the wind speed in the aftermath of a tornado is at best an educated guess. Engineering studies conducted since 1971 have shown that the wind speeds that Fujita estimated would be required to inflict the damage described in a given category were actually considerably *lower* than the speeds assigned to that category. (For instance, a well-constructed house could be leveled without the 207-mile-an-hour winds of an F4 tornado, under the original Fujita scale.) Also, the Fujita scale did not really take into consideration variations in the strength of construction, except in a general way.

An "enhanced" version of the Fujita scale, called the Enhanced Fujita scale (or EF scale), which sought to rectify these problems, was adopted across the United States on February 1, 2007. Here's how it now stands:

ENHANCED FUJITA SCALE

Category	Wind speed	Relative frequency
EF0 Light damage	65-85 mph	53.5 percent
EF1 Moderate damage	86-110 mph	31.6 percent
EF2 Considerable damage	111-135 mph	10.7 percent
EF3 Severe damage	136-165 mph	3.4 percent
EF4 Devastating damage	166-200 mph	0.7 percent
EF5 Incredible damage	> than 200 mph	< 0.1 percent

Characteristics
Peels surface off some roofs. Some damage to gutters or siding. Shallow-rooted trees pushed over. (Confirmed tornadoes with no damage, such as those that occur in empty fields, are always rated F0.)
Roofs severely stripped. Mobile homes completely destroyed. Loss of exterior doors. Windows and other glass broken.
Roofs torn off frame houses. Foundations of frame homes shifted. Large trees snapped or uprooted. Cars lifted off ground. Light objects turned into missiles.
Entire stories of well-constructed houses destroyed. Severe damage to large buildings such as shopping malls. Trains overturned. Trees debarked. Heavy cars lifted off ground and thrown. Structures with weak foundations blown away some distance.
Well-constructed houses completely leveled. Cars thrown. Small objects become missiles.
Strong frame houses lifted off their foundations and swept away. Car-size missiles fly through the air in excess of 100 yards. Trees debarked. Steel-reinforced concrete structures badly damaged. High-rise buildings have significant structural deformation.

One descriptive entry under the EF5 rating—"incredible phenomena will occur"—hardly sounds scientific. But it is a precise description of some of the devastation that follows these quite literally incredible events. (In his original scale, Fujita anticipated that categories up to F5 would cover all anticipated damage to frame buildings, but he did suggest a higher category, F6, for what he called an "inconceivable tornado." In the Enhanced Fujita scale, there is no upper limit to wind speeds considered EF5.)

It is also evident from the foregoing that tornadoes become increasingly rare as they ascend the scale of severity. More than 95 percent are EF2 or lower, and fewer than one percent are EF4 or EF5. Still, on May 4, 2007, only three months after the new scale was unveiled, tornadoes in Greensburg, Kansas, were given a rare EF5 rating.

The development of a large-scale tornado database, with its climatology map, and the Fujita scale, made it possible for researchers to see things about tornadoes that had not previously been apparent. For instance, it became clear that although the most violent tornadoes (EF3 to EF5) are comparatively rare, they are also responsible for nearly all tornado fatalities (about 88 percent). Fujita and others also pointed out that the area of maximum tornado fatality rates (east of the Mississippi, from Mississippi to Illinois) does not match the area where tornado frequency is greatest (Tornado Alley, from central Texas north to South Dakota). The difference has been attributed to greater tornado awareness, better building construction, and perhaps better visibility in the more western states.

Summing up Ted Fujita's career, University of Oklahoma meteorologist Charles Doswell says simply that "there is not much in this field that he did not touch." Fujita's many contributions, besides the Fujita scale, included these key discoveries:

- He pioneered the whole field of "photogrammatic analysis" of photos and movies of tornadoes, combined with

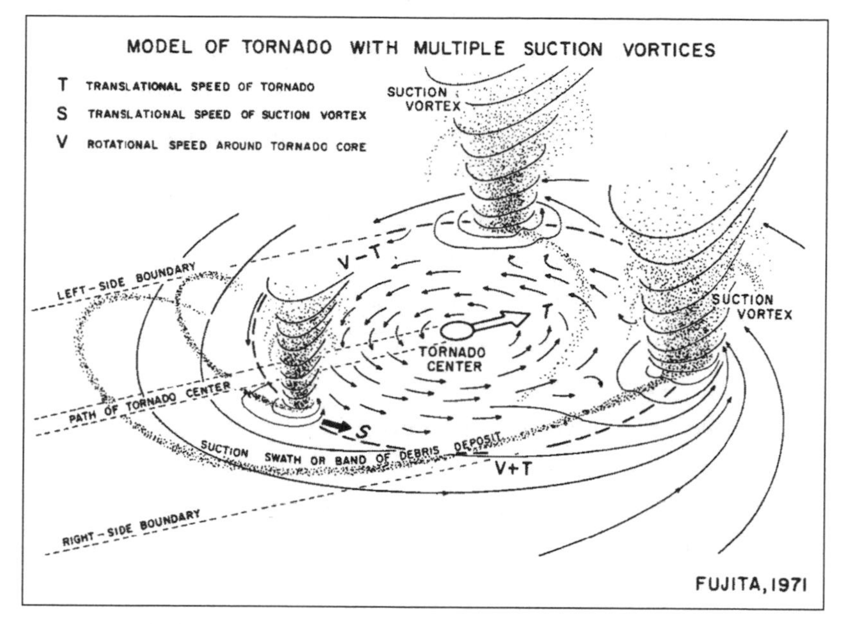

Fujita studied curving cycloidal marks in fields after tornadoes to discover that multiple suction vortices can form around the primary vortex at a tornado's core.

meticulous study of tornado damage fields. Before the advent of modern technology, especially Doppler radar (first used to measure tornado wind speeds in the late 1950s), he used deductive logic and Holmesian attention to detail to begin understanding tornadoes. "Fujita was famous for these intricate mappings, which often revealed fascinating subtornado-scale details of the tornado damage path," write meteorologists Greg Forbes and Howard Bluestein.

- He demonstrated that tornadoes sometimes contain multiple vortices, which he called suction vortices. Since the multiple tornadoes in Fargo, and even earlier, Fujita had become interested in the fact that circular or cycloidal "scratch marks" sometimes appeared on the ground after tornadoes. For instance, studying the damage after a tornado in Illinois in 1967, he found piles of corn stubble formed into this curious, circular shape, littered across the landscape (the sort of baffling clue Holmes would have loved). Fujita began referring to these circular piles as suction swaths, as if some vacuum cleaner–like structure within the tornado sucked up the debris into a circular pile but could not lift it off the surface, according to Forbes and Bluestein. Further study led Fujita to propose that the main cyclonic rotation of a tornado will sometimes have smaller suction vortices spinning around it, like supporting dancers around a ballerina, which leave these mysterious groupings as they follow the main funnel along.

- In a damage assessment of Xenia, Ohio, after an April 3, 1974, superoutbreak of tornadoes, Fujita showed that places directly *in the path of* a suction vortex could be exposed to winds as much as 90 miles an hour stronger than places *between* suction vortices. This discovery, write Forbes and Bluestein, offers "an alternative explanation to the often-asked question of why one house was spared while the one next door was destroyed." Before Fujita, the standard answer had always been that tornadoes simply "skipped" along, missing some houses and obliterating others.
- In the Northern Hemisphere, virtually all tornadoes rotate counterclockwise. But Fujita demonstrated that "anticyclonic" tornadoes, which rotate clockwise, "are not as rare, and not necessarily as weak, as once thought," according to Forbes and Bluestein.
- In 1961, while aboard a Weather Bureau surveillance aircraft, Fujita took a remarkable photograph showing a supercell thunderstorm, then produced one of his trademark diagrams to show how he thought it might work.

> He noted that there was a radar echo (indicating an area of intense turbulence or precipitation) on the southwest side of the storm. Though he did not use the term, he had just described the "hook echo" (or hook-shaped area of turbulence), which is now known to be a reliable indicator of tornadoes.

It's easy, perhaps, to forget that these scientific breakthroughs did not affect only researchers in sterile labs, or in the ivory towers of academe. They took place at the intersection between human affairs and Earth's mysterious and unstable atmosphere—the place where people live and die.

Nowhere was this more evident than in the skies over New York City on June 24, 1975. Shortly after 4 p.m. on that fateful summer afternoon, Eastern Airlines Flight 66 from New Orleans, with 124 passengers and crew aboard, ran into a ferocious electrical storm as it was approaching the runway for landing at John F. Kennedy Airport, in New York. Horrified onlookers watched as the big blue-and-white jet approached the runway at an alarmingly low altitude, clipping off the landing lights on five 35-foot towers marking the approach to the runway. The aircraft regained altitude briefly, then slammed into the ground and exploded, spewing fiery wreckage and bodies in all directions. One observer said he saw "this big flash of fire," apparently from exploding engines, like an atomic bomb, followed by flames that towered 500 or 600 feet high.

Rescuers rushed to the scene in the pouring rain, covering bodies with white plastic sheets, and tagging the odd, sad bits of debris—a shoe, a cosmetics case, a man's jacket. A couple of days after the crash, when it was discovered that a two-month-old baby who was not on the passenger manifest had been aboard, the death toll reached a total of 112. At the time, it was the deadliest airline accident in U.S. history.

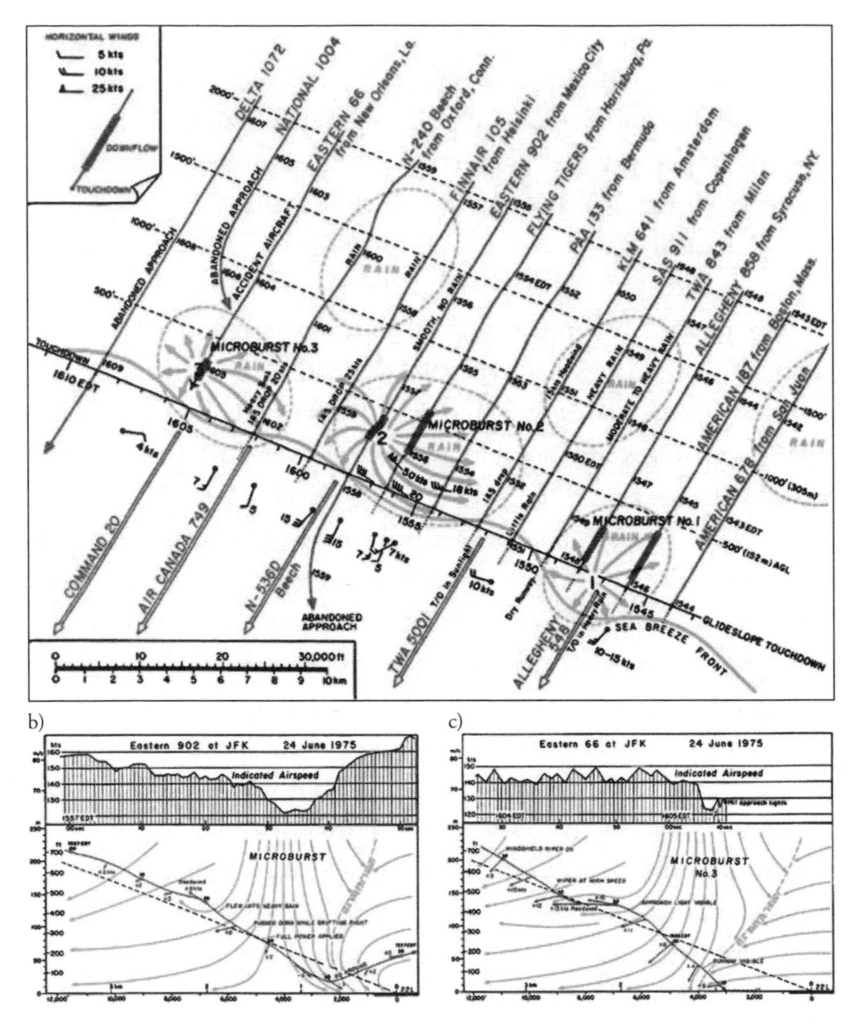

Fujita's analysis of the weather events at JFK International Airport on June 24, 1975, which caused the crash of Eastern Airlines Flight 66, led to his discovery of powerful air currents that strike the ground, called downbursts.

In the aftermath of the crash, investigators from the Federal Aviation Administration (FAA) were at a loss to explain what had happened. There was some speculation that the plane might have been struck by lightning in the storm. One woman claimed to have seen a lightning bolt surge through the jet as it dipped to within 50 feet of the ground.

But there were several things about the tragedy of Flight 66 that were deeply puzzling. For one thing, despite the storm, the overall weather conditions were well within the capability of an experienced crew—six-knot winds, a 3,000-foot cloud ceiling, and five-mile visibility. Also, there were 12 other planes that landed at JFK without incident in the same storm. On the other hand, there were also reports of severe, localized turbulence at the airport. Just before Flight 66 crashed, the captain of Eastern Airlines Flight 902, a jumbo jet, was forced to abort his landing because of violent winds. "He told his company people that he barely saved it," an official told the *New York Times.*

The FAA investigators began to focus on the possibility that wind shear had caused the crash. (That is, localized changes in wind speed or direction over a short distance, resulting in a tearing or shearing effect, which can be hazardous to airplanes, particularly if they're landing or taking off.) But if wind shear were to blame, why had so many other planes been able to land uneventfully? There was something about the mystery of Flight 66 that contemporary atmospheric science seemed unable to explain.

A safety expert named Homer Mouden, who was investigating the crash, became fascinated by the fact that Flight 66 had encountered deadly winds at almost the same time other aircraft, at the same airport, ran into nothing more than an ordinary thunderstorm. It was Mouden who sought out Ted Fujita, who by then had gained a national reputation as an expert in severe storms and tornadoes and the complex wind fields that surround them.

Fujita analyzed the aircraft flight data recorders, reports of other pilots, and wind speed data from the airport anemometer (wind speed indicator). He also remembered that, after the superoutbreak of tornadoes the previous spring, he had seen odd, small-scale damage patterns on the ground, different from the circular "suction vortices" he had discovered in the aftermath of other tornadoes.

He noticed the swirling pattern of downed timber frequently seen after tornadoes, but also curious "starburst" patterns, where trees had been ripped out of the ground by their roots and splayed outward. They were not unlike the starbursts he'd seen at Nagasaki after the atomic bomb explosion in 1945.

He proposed a working theory of what had happened to Flight 66: It had flown into a "diverging wind system," which was similar to but weaker than what caused the starbursts of destruction after the tornadoes of April 1974. He used the term "downburst" to describe both the powerful downdraft of wind and the fact that it burst outward when it struck the ground. It was the terrifying blast of the downdraft that destroyed Flight 66. In effect, the downburst was a kind of tornado in reverse, with a relatively narrow column of air descending to the ground and then spreading, instead of a broad inflow of air being sucked up into an ascending, fairly narrow column of air.

Fujita proposed dividing these hypothetical weather events into microbursts (equal to or smaller than four kilometers) and macrobursts (greater than four kilometers).

Using the scattered bits of information he had available, and pulling it all together with ingenuity and insight—his detractors would say taking a wild leap unsupported by the evidence—Fujita constructed a simple, compelling diagram of what he believed had happened to Flight 66. Crossing the page were 14 red lines, nearly straight and equally spaced apart, representing the flight paths of 14 commercial airliners that came in for a landing at JFK that afternoon. Superimposed upon these lines were three blue starburst patterns, labeled Microbursts 1, 2, and 3. Eastern Flight 902 flew almost directly into Microburst 2 and had to abandon its approach. In an accompanying diagram, Fujita superimposed the airflow of the microburst and 902's airspeed; when the two collided, the aircraft made a sickening drop in speed and very nearly plunged to zero before regaining velocity and then veering away to safety. In Fujita's diagram, ill-fated Flight 66 flew directly into Microburst 3. In a second, heartbreaking diagram, Fujita showed the plane's airspeed as it ran into the intense downburst, dropped suddenly, briefly regained altitude, and then crashed to the earth and exploded.

Ted Fujita, the Sherlock Holmes of severe storms, seemed to have solved the Mystery of Flight 66. Of course, like everything else in science, Fujita's theory would still have to survive rigorous analysis and debate.

Suction vortex A small but very intense vortex inside the circulation of a tornado; often several are present in multiple-vortex tornadoes, and are thought to cause considerable damage.

Downburst A strong downward current of air associated with intense thunderstorms that often create a star-shaped impression on the ground

In a scientific paper about Fujita's discovery of the downburst, meteorologists James W. Wilson and Roger M. Wakimoto wrote that Fujita's analysis of the weather events surrounding Flight 66 "is an excellent example of both his creativity and insight as he carefully pieced together disparate bits of data." Fujita, they went on, "rarely used computers when analyzing data, preferring to use manual analysis techniques that he had perfected early in his career. . . . Fujita's genius was in being able to take an incomplete set of observations, intuitively fitting them to real-world phenomena, and then preparing colorful, easy to understand figures."

But Fujita's downburst theory did not meet with universal acclaim, especially in the scientific community. There were those who said he had simply renamed the well-known phenomena of downdrafts and gust fronts in thunderstorms (a gust front being the leading edge of the downdraft after it hits the ground). Others said a downdraft simply could not be powerful enough to account for these tragic accidents. Fujita also initially proposed an explanation that turned out to be wrong—that downbursts were caused by the collapse of the overshooting top at the roof of a supercell.

Stung by these criticisms, Fujita set out to prove the existence of the downburst. Between 1975 and 1978 he began doing low-elevation aerial surveys to photograph starburst damage patterns in cornfields and forests after severe storms and tornadoes. But this was only circumstantial evidence; he needed direct evidence of airflow patterns to solve the case to the satisfaction of his critics. In the following years, Fujita became involved in two field programs, dubbed NIMROD (Northern Illinois Research on Meteorological Downbursts) and JAWS (Joint Airport Weather Studies), in which multiple Doppler radar sites and other advanced technology were used to prove once and for all the existence of the downburst and how it worked. Ultimately, his theory was confirmed.

Fujita was "the scientific genius behind the discovery of the convective weather phenomenon called the downburst," Wilson and Wakimoto say in their paper. The "transfer of this knowledge into the aviation community [has] benefited the whole of society and must be considered one of the major, rapid payoff, success stories in the atmospheric sciences."

By 1982, Ted Fujita, now 61, had become known to the world as Mister Tornado and was considered the world's premier tornado and severe-storm expert. Yet, incredibly enough, he had never actually seen a tornado himself.

That is, until June 12, 1982. That afternoon, Fujita and a group of scientists were visiting one of the radar sites for the JAWS program, hunting for more data about downbursts west of Denver. A line of cumulus clouds was forming to the east, billowing up into storm towers in the golden late-afternoon light. The team had three small Doppler radar dishes, mounted on flatbed trailers, pointed toward the storm clouds. Rita Roberts, a scientist with the National Center for Atmospheric Research (NCAR), recalled later that Fujita kept running in and out of the trailer, snapping pictures of the towering thunderhead while synchronizing each picture with his watch.

Then, to the amazement of everyone—especially Fujita—not one but two beautiful tornadoes appeared out of the distant cloud, cool and white and perfectly lit for photography. Everybody cheered and joined Fujita taking pictures.

For years, Fujita had driven around with a license plate that read "TF0000," referring to his initials and the number of tornado viewings he had experienced. Now, the greatest tornado scientist in the world told the gathered researchers, he'd finally have the right to change his license plate to "TF0001." In the end, though, he never changed his license plate, using TF0000 until the end.

11: A LIFE-THREATENING SITUATION

ON THE AFTERNOON OF JUNE 16, 1928, A WHEAT FARMER FROM GREENSBURG, KANSAS, named Will Keller was out sadly inspecting the ruins of his crop after it had been completely beaten to the ground by a ferocious, passing hailstorm.

Catastrophe was common in Kansas in the 1920s: Though the Great Depression and the Dust Bowl years were yet to come, Keller had lived through enough tornadoes to recognize that the eerie, umbrella-shaped cloud he now spotted in the sky southwest of his farm very likely contained a twister. He also noticed "that peculiar oppressiveness which nearly always precedes the coming of a tornado," in the words of Alonzo A. Justice, an employee of the Weather Bureau office in Dodge City, Kansas, to whom Keller later told his story.

Keller turned his attention back to surveying the damage to his farm, but when he glanced back at the umbrella cloud, he saw that "hanging from the greenish-black base of the cloud was not just one tornado, but *three.*" One of the tornadoes was already "perilously near and apparently headed directly for our place," so he lost no time hustling his family down into the cyclone cellar for safety.

Having lived through many tornadoes out on the Great Plains, Keller calmly paused in the doorway of the storm cellar to watch what happened next. The tornado—or rather, tornadoes—was "indeed an impressive sight. . . . Two of the tornadoes were some distance away and looked to me like great ropes dangling from the clouds, but the near one was shaped more like a funnel with ragged clouds surrounding it. It appeared to be much larger and more energetic than the others, and it occupied the central position of the cloud, the great cumulus dome being directly over it."

As he boldly watched, Keller noticed that the lower end of the funnel, which had been sweeping across the ground, appeared to rise. "I knew what that meant, so I kept my position. I knew that I was comparatively safe and I knew that if the tornado again dipped I could drop down and close the door before any harm could be done. . . .

"At last the great shaggy end of the funnel hung directly overhead. Everything was as still as death. There was a strong gassy odor and it seemed that I could not breathe. There was a screaming, hissing sound coming directly from the end of the funnel. I looked up and to my astonishment I saw right up into the heart of the tornado. There was a circular opening in the center of the funnel, about 50 or 100 feet in diameter, and extending straight upward for a distance of at least one half mile, as best I could judge under the circumstances. The walls of this opening were of rotating clouds and the whole was made brilliantly visible by constant flashes of lightning which zigzagged from side to side. Had it not been for the lightning I could not have seen the opening, not any distance up into it anyway.

"Around the lower rim of the great vortex, small tornadoes were constantly forming and breaking away. These looked like tails as they writhed their way around the end of the funnel. It was these that made the hissing noise. I noticed that the direction of rotation of the great whirl was anticlockwise, but the small twisters rotated both ways—some one way and some another.

A LIFE-THREATENING SITUATION

"The opening was entirely hollow except for something which I could not exactly make out, but suppose that it was a detached wind cloud. This thing was in the center and was moving up and down."

After giving Will Keller this breathtaking peek into its mysterious core, the tornado roared away. It touched down again a few miles to the northeast, destroying the farm of a neighbor named Evans. Evans did not have time to shepherd his family into the cyclone cellar, so they dove behind a small bluff out of the wind. Evans later said he looked up and saw the family cook stove revolving around and around high up in the air; moments later, the cyclone stripped every bit of clothing off his 17-year-old daughter's body, though she was otherwise unhurt, along with the rest of her family.

Hazardous Weather Outlook A statement issued by the National Weather Service (NWS) and updated regularly regarding the potential of significant weather expected over the next day to five days

Thermodynamics The relationship between heat and other properties, such as temperature, pressure, and density; in meteorology, it most often refers to temperature and moisture.

Seventy-nine years later, long after the Depression, the Dust Bowl, a multitude of wars, the Internet, and everything else that has happened in human affairs, the likelihood of tornadoes in Kansas has hardly changed at all. In fact, between January 1950 and March 2007, at least 25 tornadoes touched down in Kiowa County, the south-central Kansas county where Greensburg (and Will Keller's farm) is located, according to a paper by William Monfredo, of the Applied Disaster and Emergency Studies Department at Brandon University. Still, all but three of these tornadoes were weak, and no one had been killed by tornadoes in the county in more than half a century.

That is, until darkness fell on the terrible evening of May 4, 2007.

It had been apparent that something potentially dangerous was developing in the atmosphere over the central Plains states as much as a week earlier. On April 27, the National Weather Service (NWS) office in Dodge City, Kansas, about 40 miles from Greensburg, had issued a Hazardous Weather Outlook, anticipating a threat of severe weather over the coming week.

As the week wore on, several things caught the attention of meteorologists in the Dodge City office, including a 27-year-old forecasting meteorologist named Mike Umscheid.

The atmospheric developments also caught the attention of Tim Samaras and his team, who began planning a mission into the area later in the week from home base in Denver. Carl Young, the whip-smart meteorologist who often rode shotgun with Tim, flew in from his home in California the night of May 3. The next morning, Tim, Carl, and a small team of students and researchers assembled in a motel parking lot in Denver.

"Even though we knew something significant was developing, it's difficult to see long-term patterns evolving that would show exactly what section of the country was going to get hit by tornadoes," Carl says. "Tim and I felt like Oklahoma and Kansas were going to be under the gun. But partly because we got a late start that morning, making it to the southern target, around Greensburg—350 miles or more—seemed like quite a push. Also, we wanted a target that would likely fire during the daylight hours, and there was a concern that the southern target would be a nighttime event."

So Tim and the TWISTEX team decided to head east on I-70, chasing the huge, roiling storm cell developing in west-central Kansas a couple of hours' drive north of Greensburg.

There are basically four ingredients for creating severe storms and tornadoes, Mike Umscheid explains. One is low-level moisture, or wet air close to the ground—the jet fuel that powers storms. The second is wind shear, or the winds that tend to spiral as they gain elevation, thus increasing the likelihood that a storm will begin to rotate. The third is atmospheric instability, which essentially means that air masses will rise rapidly rather than being stalled at a lower elevation. And the last ingredient is some sort of triggering mechanism—the spark dropped in the fireworks factory. When something causes an ordinary summer thunderstorm to "initiate" or "fire," and all the other elements are in place, the result can be absolutely explosive. All four of these elements were falling into place—in fact, they were "maxing out"—in the days before May 4, Umscheid noticed. That was highly unusual.

On the morning of May 1, the Storm Prediction Center (SPC), a bureau of the National Weather Service in Norman, Oklahoma, issued a four-day outlook highlighting the threat of severe weather in the central Plains states. When the more short-term forecasts were issued, they also looked increasingly ominous.

The SPC is able to analyze a database of historical data to find similar thermodynamic profiles. (The term thermodynamics refers to the mechanics of heat in the atmosphere.) Now this model was showing something very worrisome indeed: The developing storm system matched the profiles of such cases as May 3, 1999. That was bad—in fact, it was very, very bad. May 3, 1999, was the date of the disastrous F5 tornado outbreak in Moore/Oklahoma City, Oklahoma—it was, in fact, the first F5 outbreak in eight years. This horrendous outbreak lasted three days, resulted in more than 60 tornadoes, killed 38 people, and caused almost $1.5 billion in damage.

In the most significant of these tornadoes, an F5 that crossed into Bridge Creek, Oklahoma (killing 36 people and demolishing more than 10,500 buildings), a Doppler on Wheels radar dish system detected wind speeds of an incredible 301 miles an hour. Since this number was accurate only to within plus or minus 20 mph, it was actually conceivable that the wind speeds were as high as 321 mph. This wind speed was so high, in fact—at the extreme high end of the highest category in the Fujita scale—that it led to speculation that a new category, F6, might need to be added to the scale.

(In fact, in several recent papers, aeronautical engineer W. Steve Lewellen of West Virginia University has suggested that it is possible maximum wind speeds inside the tornadic core could briefly approach the speed of sound—that is, 767 miles an hour or more. The reason such wind velocities have not been documented so far, Lewellen argues, is that they are hidden in the core by heavy clouds of debris; the winds are mostly vertical, the most difficult kind of wind for Doppler radar to measure; and they occur in the lowest 100 meters of the tornado, which is extremely difficult and dangerous to study—and which happens to be the same region that has been the focus of Tim Samaras's research. Although still unproved, the fact that such nearly supernatural wind velocities are under serious scientific discussion shows just how much more about tornadoes remains to be discovered.)

Another forecasting tool, called the Significant Tornado Parameter, was also going off like the alarm from a five-alarm fire. This tool, which produces a single number to represent a variety of severe weather conditions like wind shear and CAPE (convective available potential energy), was registering a 7, where a parameter value of 1 meant blue sky, daisies, and utter atmospheric boredom, and anything above 1 showed an increasing chance of supercells, tornadoes, and atmospheric terror.

As Umscheid watched the whole "synoptic" (large-scale) system develop, he planned his next steps: "When I see something like this, I'm trying to figure out how I can get off work." This comment was met with huge laughter at a 2008 storm chasers convention in Denver (an event that was born in Tim Samaras's basement), where the hall was filled with people who had manufactured all kinds of excuses to get time off to go chase storms. Though he was a professional forecasting meteorologist with the National Weather Service (NWS), Umscheid was also an avid part-time storm chaser and photographer. Still, at a certain point, the prospects for a rollicking good storm begin to darken into genuine apprehension that something terrible is about to happen, and that innocent people could be in grave danger. That's what seemed to be happening now.

A LIFE-THREATENING SITUATION

Tim was lucky enough to be off work for two whole months (arguably, he'd been "off work" for his whole life). He had a generous and flexible arrangement with his employer, the engineering firm Applied Research Associates, so that he was free to go storm chasing for the months of May and June. Now, he and his team headed east on I-70 as the skies began to darken and tornado warnings were issued by the NWS for the area they were targeting.

Early May is the heart of severe-weather season in western Kansas. When the jet stream, moving through the area, begins to form a trough, or a long area of low pressure, and especially when there are southwest winds in the jet stream, a huge atmospheric turbulence is created, which tends to produce thunderstorms:

- Tornado ingredient number 1. Now a trough was moving toward western Kansas, and this weather feature persisted day after day. There was also "classic, classic southwest flow" into the central High Plains, Umscheid noticed—just the sort of inflow that often led to immense storms. The "dry line," an unstable boundary between warm, moist air coming up from the Gulf and dry air flowing down off the Rockies to the west, was also moving into western Kansas. If there were going to be supercells or tornadoes, they would probably form just to the east of the dry line. It was as if a bull's-eye were forming somewhere in southwestern Kansas.
- Tornado ingredient number 2: There was plenty of "jet fuel" to stoke a major storm, too. The surface dew points ranged from the low to middle 60s across central Kansas, to around 70 degrees across central Oklahoma. Dew points greater than 60 degrees are best for severe-storm formation; the atmosphere is said to have

"juice" when dew points exceed 60 degrees. The atmosphere was now absolutely loaded with juice. If a storm got started, it could feed on all that fuel for hours, or even days.

- Tornado ingredient number 3: There was also a tremendous amount of upward vertical motion in the atmosphere, as all this warm, soggy air evaporated. At the same time there was plenty of wind shear, tending to nudge the whole thing over into the gradually accelerating rotation that could turn into a gigantic aerial whirlpool. As National Weather Service meteorologist David Floyd says, "rotation significantly enhances the vertical velocity of the updraft." In other words, the faster the spin, the faster air gets sucked aloft.

It was as if the atmosphere were just dying to make a tornado. All it needed, it seemed, was ingredient number 4: the spark.

Even so, all this atmospheric turbulence—this potential for chaos—was capped, meaning that there was a half-mile-thick layer of warm, dry air lying on top of the whole thing, roughly 7,000 feet above the ground. This layer of warm air (called temperature inversion) was like a lid on a pot of boiling water. Nothing could "fire" until something happened to break through the cap. When that happened, all bets were off—and human life and property for hundreds of miles in every direction were at risk.

"This temperature inversion ensured sun-filled, rain-free skies throughout the area for much of the daylight hours," observed severe-weather expert William Monfredo. But the sunny blue skies were

GLOSSARY

Jet stream A narrow stream of strong, horizontal, high-altitude winds, which varies from day to day

Cap A layer of warm air, usually several thousand feet above the ground, which delays the development of thunderstorms and is an essential ingredient to severe-weather episodes

Jet streak The area of maximum wind speed inside a jet stream

gravely misleading: Invisible to the unpracticed eye, terrible danger was gathering. The cap kept an increasingly violent and turbulent atmosphere damped down until late in the afternoon, when the cap finally began to break and supercells started geysering up to the edge of the stratosphere, potent and terrible.

Mike Umscheid noticed it too: a small area of enhanced heating in the eastern Oklahoma Panhandle. "This, I think, was the key to the initiation process [ingredient number 4]," he said later—the spark that set the bonfire ablaze. "At 2 p.m. that afternoon of May 4, there were places in Oklahoma that were approaching 90 degrees. Ninety degrees will almost always remove the cap. If you're getting 70s or low 80s, that's not going to cut it as far as breaking the cap. But 90 will almost always do it."

In addition to the four tornado ingredients, something else was also happening: A jet streak had begun to form—a kind of tongue of higher wind velocities embedded within the jet stream, usually a few hundred miles long, that can help to erode the cap and, if it passes over the top of an unstable air mass, release explosive storms. That's what appeared to be happening now. The lid that was keeping everything bottled up was growing flimsier by the minute.

Incorporating all of this information into its daily "Day One" report, the SPC issued a Moderate Risk of severe storms with a 15 percent probability for tornadoes, and at least a 10 percent probability of tornadoes that could produce EF2 to EF5 damage. In other words, there was a betting chance that this day would produce not just tornadoes, the most violent windstorms on Earth, but the most severe kind of tornado, a very rare EF5.

Unable to get off work that day to go storm chasing, Umscheid reported to the radar desk at the Dodge City office for the 2 p.m. to 10 p.m. shift. This was often called the radar shift, because most of the severe weather occurred in the late afternoon and early evening; the radar shift operator was the person tasked with monitoring storms on radar and getting the word out if things got ugly. Thus Mike Umscheid got a ringside seat into what he later called "the most interesting and unusal severe weather event I have ever seen.

"I am fascinated by supercell thunderstorms—but you've got ordinary supercell thunderstorms, and then you've got storms like what happened May 4, 2007, in Greensburg, Kansas. Storms that are complete anomalies and are just absolutely fascinating in every facet."

At 6:15 p.m., the Storm Prediction Center issued a tornado watch for parts of western-central Kansas, northwestern Oklahoma, and the eastern Texas Pandhandle:

> **EFFECTIVE THIS FRIDAY NIGHT AND SATURDAY MORNING FROM 615 PM UNTIL 200 AM CDT . . . TORNADOES . . . HAIL UP TO 3 INCHES IN DIAMETER . . . THUNDERSTORM WIND GUSTS TO 70 MPH . . . WIDELY SCATTERED STORMS /SUPERCELLS POSSIBLE IN VERY UNSTABLE AIR MASS ALONG SLOWLY RETREATING DRY LINE THIS EVENING . . . STRENGTHENING ASCENT/ WIND FIELD . . . MAY ENHANCE TORNADO POTENTIAL AFTER DARK AS LOW LEVEL JET INCREASES TO AROUND 50 [KNOTS].**

It was an ominous warning, but not one that betrayed the true magnitude of what was about to happen.

By this time, Tim's team was bearing down on the immense storm cell along I-70. "We were seeing huge cumulonimbus storm towers going up" in the late afternoon light, Carl says. This cell looked so promising that Josh Wurman's team, with his truck-mounted Doppler radar dishes, the preposterous-looking Tornado Intercept Vehicle, and a caravan of TV production people, had come north to chase the same cell the TWISTEX team was chasing. Another prominent tornado meteorologist, Howard Bluestein, stayed to cover the Greensburg storm only because he had a flat tire and got stuck there. "Storm chasing involves a whole lot of luck," Tim says.

A LIFE-THREATENING SITUATION

Storm chaser Warren Faidley, positioned on a Kansas hilltop monitoring radar and SPC bulletins, watched immense storm towers going up in the golden early evening light to his northeast (around Greensburg) and also to his southeast (down by the Texas/Oklahoma border). The storm tower to the northeast, he recalled later in an article, "rises upward and fans out in the likeness of a dragon's head with its mouth open." Faidley had to decide quickly which of the two unfolding storms he would chase. He decided to go after the southern storm, near Arnett, Oklahoma, along with many other storm chasers. That produced an amazing tornado—but nothing like what was about to happen in Greensburg.

Like almost every other little town on the Great Plains, Greensburg, Kansas, boasted that it was the home of the biggest, longest, tallest, or oldest something or other. In Greensburg's case it was the "World's Deepest Hand-Dug Well," all of 32 feet in diameter and 109 feet deep. Completed in 1888, it had provided the town's water supply until 1932. The town was so proud of it that an enormous billboard at the edge of town announced, "Greensburg has a great BIG WELL-come for YOU!" The other curiosity on display in the town, much less publicized, was a thousand-pound meteorite—proof that, like tornadoes, alien, dangerous things could sometimes fall out of the prairie sky with almost no warning at all.

Greensburg was a town of about 1,500 people. The median value of a single-family home, $46,000, was less than half the national average; the percentage of residents older than 65 was more than twice the national average. It was one of those sweet, small, left-behind midwestern towns, steadily losing population as the young grew up and departed for places with more jobs and more excitement.

The first tornado was reported to the SPC by a storm chaser near Clark, Kansas, at 7:40 p.m. The storm cell that eventually became the Greensburg tornado developed over north-central Harper County in northwest Oklahoma at about 7:50 p.m., moving north-northeast.

The first ominous hook echo appeared on Umscheid's radar screen at 8:06 p.m. over south-central Clark County. At 8:13 p.m., the Dodge City office issued its first tornado warning, for Clark and Comanche Counties. The radar showed a powerful mid-level mesocyclone, which persisted as it moved into Kiowa County. Then it rapidly developed a second hook signature, this time much more powerful than the first one.

Another chaser reported, via cell phone, a rope tornado near Comanche shortly after 8 p.m., followed moments later by another report of a tornado on the ground with debris clouds, also in Comanche.

Then, around 9 p.m., there were reports of an entire "family" of tornadoes—an immense wedge tornado with a satellite tornado and a rope tornado, all rapidly becoming "rain-wrapped"—disappearing into the chaos and confusion of the supercell, so that it was impossible to tell exactly where they were. It was getting so dark that storm chasers had to rely on lightning and "power flashes" from collapsing utility lines to see what was going on. It was an extremely dangerous situation. Digital photographs and video frame grabs snatched from these moments show an eerie, immense, black wedge tornado hanging down out of the ceiling of the wall cloud, with twilight glimmering just above the horizon line.

The I-70 storm cell Tim's group was following did produce a couple of small, weak tornadoes, but "nothing of any significance," Carl says. They were unable to deploy a probe. "Frankly, it was kind of disappointing, but that's what happens sometimes," Carl says. And because it was beginning to get dark, they'd begun to think about calling it quits. Then the radar images from down around Greensburg, and the bulletins from the Dodge City office of the National Weather Service, became increasingly alarming.

"This storm had all the classic tornado signatures, including a very well-defined, pronounced, and persistent hook," Carl says. "We were not only getting radar on Threat Net, through XM Satellite Radio, but also much higher resolution radar coverage, through alternative Internet sources, which showed an amazing, powerful storm, a wicked storm. And we could tell from the frequency and content of the warnings, which were coming out every few minutes, that those people down there were in serious danger."

It was Mike Umscheid, in the Dodge City office, who was issuing most of those warnings. At 9:19, Umscheid issued a tornado warning for Kiowa County:

> **BULLETIN—TORNADO WARNING—THE NATIONAL WEATHER SERVICE IN DODGE CITY HAS ISSUED A TORNADO WARNING FOR KIOWA COUNTY IN SOUTH CENTRAL KANSAS—AT 9:17 PM NATIONAL WEATHER SERVICE METEOROLOGISTS WERE TRACKING A CONFIRMED LARGE AND EXTREMELY DANGEROUS TORNADO 14 MILES SOUTH OF GREENSBURG MOVING NORTHEAST AT 25 MPH—THIS IS AN EXTREMELY DANGEROUS AND LIFE THREATENING SITUATION—TAKE COVER IMMEDIATELY IN A BASEMENT OR OTHER UNDERGROUND SHELTER AND GET UNDER SOMETHING STURDY.**

Mike Umscheid could hardly believe what he was seeing on the radar screen. The data coming in were "just incredible. I have never seen 80-plus knots of tangential velocity associated with a TVS [tornado vortex signature]. There was just a phenomenal amount of wind shear and rotation. I have never been so anxious in my life. I knew there must be a very substantial amount of damage out there. But I couldn't get carried away by emotions, because my job was to relay accurate information to all the people in the surrounding area."

So he just focused on the stream of incoming data. Now he saw something that almost made his jaw drop. On the radar he could see a vivid, doughnut-shaped image—a low-level debris field that had been centrifuged by the tornado into a ring-shaped killing field. It was, Monfredo would later say, "essentially a town blowing away." Anyone or anything trapped inside that rotating ring, filled with flying missiles—not just huge hailstones but bits of sheet metal, road signs, garbage cans, and anything else that the storm could wrench free—would probably be badly injured or killed.

When you get an EF5 bearing down on your location, it's time to get somewhere belowground. "There's really no structure aboveground that's going to survive a tornado like that," says Larry Ruthi, another meteorologist in the Dodge City branch. But that's still not enough: You can still be injured or killed by debris when you're cowering in a roofless basement and something falls on top of you, so you need to get underneath something like a heavy table.

By 9:29, as the tornado approached Route 183 outside of Greensburg, it had morphed into "a gigantic, earth-grinding, one-mile-wide-plus monster," Umscheid says. And it appeared to be heading straight for the town.

Umscheid now became convinced that, "given what the data was showing, there was a 90 percent chance that something catastrophic was about to happen to Greensburg. This was not just the real deal, it was more than the real deal."

Even though his previous tornado warning was strongly worded, he now felt that something even more urgent was needed. After a 30-second consultation with another meteorologist, Umscheid issued the strongest possible warning, a rarely used "tornado emergency" alert. The time was 9:41 p.m.

"I knew I had to use enhanced wording, to really pull out the stops, because sometimes that's what it takes to get people to recognize the danger and take action."

SEVERE WEATHER STATEMENT—**A TORNADO EMERGENCY FOR GREENSBURG—NATIONAL WEATHER SERVICE METEOROLOGISTS AND STORM SPOTTERS WERE TRACKING A LARGE AND EXTREMELY DANGEROUS TORNADO—THIS TORNADO WAS LOCATED 5 MILES SOUTH OF**

A LIFE-THREATENING SITUATION

GREENSBURG . . . MOVING NORTH AT 20 MPH—A VIOLENT TORNADO WAS ON A DIRECT PATH FOR PORTIONS OF GREENSBURG . . . ESPECIALLY THE EASTERN PORTIONS OF TOWN. TAKE IMMEDIATE TORNADO PRECAUTIONS—THIS IS AN EMERGENCY SITUATION FOR GREENSBURG!!

A high school student named Megan, whose family lived directly across the street from Greensburg High School, was at home when she heard local TV meteorologist Dave Freeman say that the tornado was gunning directly for Greensburg and would probably hit the town about 9:52 p.m. Outside the house she heard the tornado sirens wailing. Rain and hail lashed the windows. The wind shrieked in the eaves. Suddenly the power went out. She threw herself down on the living room floor with her face buried in the couch.

"My ears started popping really bad," she recalled later. "I mean, it was worse than going up in an airplane really fast. It just plain *hurt.*"

Humans' inner ears are extremely sensitive to pressure changes. What Megan was experiencing was an extreme drop in barometric pressure, meaning that she was in or near the center of the tornado. Tim Samaras's 2003 probe data from the Manchester, South Dakota, tornado showed an incredible 100-millibar pressure drop in a matter of seconds. (By comparison, riding the elevator to the top of the 1,300-foot Sears Tower, in Chicago, results in a 57-millibar pressure drop.)

In a house nearby, wheelchair-bound Frank Gallant had no place to go, because he had no basement to flee to. So he wheeled his chair to the center of his house and hoped for the best. "You just hope you've lived up to the Lord's expectations, and you're going to go to the good place and not the bad," he told a reporter later.

A truck driver named Joe Peraza heard the tornado sirens and sought shelter in the safest place he could find—the walk-in freezer at a convenience store. He and several others hid inside while the tornado tore through the town. The tornado ripped off the back side of the freezer, but Peraza made it through. When he went to look for his truck, loaded with 40,000 pounds of oil, it had been tossed aside "like nothing."

As storm chasers and spotters began calling in reports, via cell phone or e-mail, the SPC started reporting alarming details about what had just happened:

> **10 FATALITIES, 63 INJURIES . . . OF 9 PERSONS TRANSPORTED TO WESTERN PLAINS HOSPITAL, DODGE CITY, ONE WAS DEAD ON ARRIVAL, 55 OTHERS TRANSPORTED TO PRATT HOSPITAL . . . ONE FATALITY, ONE INJURY, HUSBAND KILLED AND WIFE INJURED WHEN HOUSE WAS DESTROYED . . . 2 HOUSES COMPLETELY DESTROYED. ONE GONE FROM ITS FOUNDATION AND NO REMAINS FROM THE HOUSE.**

Mike Umscheid's tornado emergency warning may well have frightened some people enough to save their lives. In fact, meteorologist Mike Smith later estimated that it could very well have saved more than 200 lives. Why? Because the Greensburg tornado was eerily similar to the deadliest tornado in Kansas history 52 years earlier. The tiny town of Udall, population 500, was struck by a savage tornado at 10:35 p.m. on May 25, 1955, killing 82 people. Mike Smith discovered "uncanny" similarities between the two storms. The Udall tornado was classified as an F5, and Greensburg, under the new classification system, as an EF5. Both destroyed 95 percent of the towns they struck and damaged the remaining 5 percent. Both traveled north, rather than making the usual southwest-to-southeast track. Both were hard to see, being wrapped in rain and hail at night. Both struck at essentially the same level of darkness. And the radar echoes of the supercell thunderstorm complexes that produced both storms were so similar that when Smith laid one map atop the other, they matched almost precisely.

The reason the death toll in Udall was so much higher—"only" ten people were killed in Greensburg—was that much of the town was asleep when the huge tornado touched down, less than half an hour after television forecasters erroneously announced that the threat of severe weather was gone.

Umscheid is not boastful about his role in warning the residents of Greensburg. The early and accurate predictions were "mainly due to advances in technology," he says. "I'm just glad to be part of the system."

When the Greensburg tornado was rated an EF5 by the National Weather Service, it was the first time any tornado had been rated a 5 since the May 3, 1999, twister in central Oklahoma.

The Greensburg tornado was on the ground for 22 miles nonstop and lasted 30 minutes. Maximum winds were estimated at between 200 and 210 miles an hour. It was the first fatal storm in Kansas in 40 years. One of the odd and eerie things about the tornado was that it made a 270-degree turn and "almost reattacked Greensburg," according to one meteorologist.

One of the reasons this tornado was given an EF5 rating was an examination of the local high school. Built in 1939, the school had exterior walls made of two courses of brick and mortar, interior walls made of concrete block and mortar, and three-quarter-inch-thick plaster on the inside walls. Yet the tornado ripped down most of these walls like so much cardboard. The elementary school nearby also sustained major damage, with only part of one wall remaining standing.

A large grain-storage bin, filled to capacity with grain and weighing an estimated 25 tons, was tossed two blocks and came to rest near the trunks of a row of snapped-off, debarked trees. There were also many homes "swept clean with little trace of debris." They were simply gone; even the crumbs were gone.

Was there anyplace safe to go, if you were trapped in the middle of this thing? Monfredo, whose special interest was in damage and survival, observed that "although badly damaged in some instances, inner hallways and interior rooms occasionally stood when outer walls did not. Plumbing added additional strength to bathrooms, allowing a few to ride out the storm in their bathtubs. All ten people who died had not sought shelter belowground." Yet even belowground shelter was no perfect refuge: One man was injured (though not killed) when a truck fell into the roofless basement where he was cowering, trapping him between the vehicle and a couch.

One of the things that amazed Mike Umscheid most about the Greensburg storm was the fact that this was not the end of the story. The extraordinarily powerful supercell thunderstorm that gave birth to the EF5 that devastated the town just kept on spawning tornadoes, one right after the other, between 10 p.m. and shortly after midnight. "This thing was just one giant washing machine," he said. A total of 18 tornadoes were reported in the Dodge City forecast area that night, with 47 tornado reports in Kansas, Nebraska, and Missouri. Most were relatively weak, but four of them (including Greensburg) were monsters.

As the Greensburg tornado dissipated, a second tornado formed almost immediately—what's known as a tornado hand-off occurred, in which two tornadoes were briefly on the ground simultaneously, one dissipating, one being formed. The second tornado, an EF3, touched down at 10:02, traveling on the ground for 21 miles, lasting an hour. The damage area was 35 square miles. This tornado, to Umscheid's amazement, was actually *bigger* than the Greensburg tornado. "I have never seen such a strong cyclonic signature from a supercell thunderstorm—it was just mind-boggling."

A third tornado formed at 10:40, northeast of Haviland, traveling 17 miles, killing one man in Pratt County. This one was an EF3, duration almost an hour, path length 18 miles, mean width about 1 mile—damage area 15 square miles.

The *fourth* tornado touched down at 11:25 p.m., two miles southeast of the tiny town of Macksville. In pitch darkness, it traveled on the ground for 13 miles, mostly through fields and rural areas. But though this last tornado was later rated at "only" an EF2 (with wind speeds up to 135 mph), it added a tragic epilogue to the Greensburg story. Stafford County sheriff's deputy and storm spotter Tim Buckman, 46, was apparently overtaken by the tornado in the dark, and his vehicle was thrown off the road. Badly brain injured, he was transported to a hospital in Wichita and put on life support. It was there in the hospital room, a couple of days later, that his 18-year-old daughter Kylee arranged to be married. A few hours after the service, officer Buckman died.

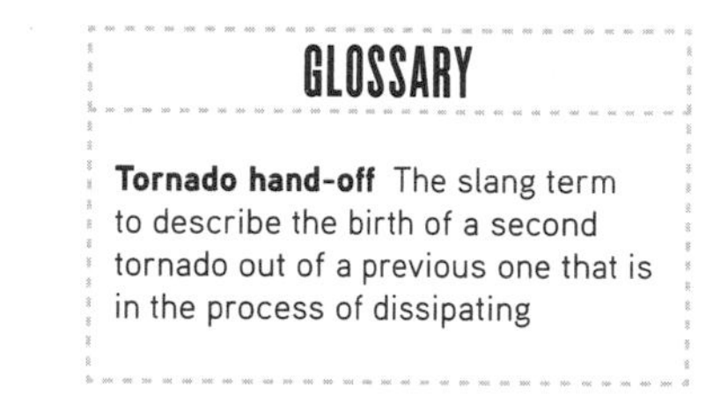

GLOSSARY

Tornado hand-off The slang term to describe the birth of a second tornado out of a previous one that is in the process of dissipating

"He died being a hero," his son Derick told a local newspaper. "He was sworn to protect the people, and that's what he was doing the night he got picked up by a tornado."

The day after what Mike Umscheid described as a "gigantic, earth-grinding monster" rampaged through Greensburg, Tim and Carl drove down to Greensburg to try to understand what had happened. The entire town was closed to the public by law enforcement. But even from a mile away, they could see that a terrible tragedy had occurred. Virtually every building was gone, except for the concrete, ten-story grain elevator, towering over the town like a gravestone.

Six weeks later, storm chaser Matt Biddle, photographer Carsten Peter, and I drove to Greensburg, which was in the process of being bulldozed to the ground. A visit to this ruined place seems to silence all comment. One feels like whispering, as if this were a church or an open grave, which it is.

Unlike hurricane damage, which becomes obvious a hundred miles or more before you arrive at the scene of greatest devastation, tornado damage tends to be highly focused, a dancing laser of destruction. You are almost upon the town before you notice anything unusual at all.

Then you can see it: On the horizon line, a collection of spindly sticks pointing jaggedly at the sky—what once were trees, perhaps handsome shade trees, now almost completely shorn of branches, leaves, and sometimes even bark. A town that grows up on the prairie begins to feel like home only when the shade trees mature, casting circles of coolness and shadow along the streets and yards. But now the trees are all gone; there is hardly a postage stamp of shade to shelter Greensburg from the baking prairie sun.

Ninety-five percent of the town has been destroyed—virtually everything. Even the half-moon-shaped debris fields that Ted Fujita sometimes noticed after tornadoes, artfully dispersed by "suction vortices," are nowhere in evidence; everything has been swept bare by the devouring wind.

The only structures that emerged intact are the grain elevator, City Hall, and the local liquor store. The fact that the Liquor Lodge ("Budweiser Welcomes You") emerged almost unscathed while every church in town was razed to the ground is one of those metaphysical conundrums for which tornadoes are renowned.

It is an eerie, heartbreaking scene. Most houses have been obliterated; others, teetering perilously on ruined foundations, are facing the bulldozer.

Only by asking directions can a visitor find the way to the World's Deepest Hand-Dug Well, because there are no signs; in fact, there are no street signs, no houses, and barely any streets at all. The entrance to the well, once a minor tourist attraction complete with color brochures and little souvenir plates, is surrounded by chainlink fence. It is forlorn and unwelcoming in the blazing sun.

This deep well, cool and dark, which once provided sweet water and sustenance out here on the prairie, is a collapsed ruin. Piles of soggy brochures and broken souvenir plates are scattered about. Nearby, a water tower said to have contained 55,000 gallons of water has been completely destroyed. Not far away, a galvanized steel grain elevator has been twisted into contemporary art, like a fantastic Frank Gehry creation.

A demolition crew is taking down what remains of the Pleasant View Motel, which has had its roof peeled off like a tin can, with the strangely composed interiors still in place—the bed, the bedside table, the chairs, the picture on the wall depicting an idyllic prairie scene of a small homestead, grazing cattle, and a windmill. There's even a wristwatch sitting neatly on a desk, as if its owner had lain down for a nap and would soon awaken to claim it.

There is a strange mix of comforting, familiar summer sounds in the air—the sounds of nighthawks and chimney swifts tittering in the high air, the sound of crickets, the sound of wind soughing through long grass. These are the sort of sounds that an astronaut in space might recall with the fiercest sort of longing, remembering the terrible sweetness of life on Earth. This is overlaid with an eerie cacophony of other, more unsettling sounds, sounds from a world hostile to humans. There is the discordant screech of a piece of twisted tin caught in a tree and moving in the wind. The creak of broken houses, badly listing and about to collapse. And then there's the eerie absence of sound—no children playing, no cars passing by, no distant radios or laughter or sounds of human life at all except the growl of front-end loaders, knocking down what is left of the town.

12: CALLING ALL TORNADO CHASERS

IT WAS, AT THE TIME, ONE OF THE MOST ELABORATE AND EXPENSIVE SPECIAL EFFECTS in Hollywood history. The cost alone was astonishing: A steel gantry custom built by Bethlehem Steel cost a mind-boggling $12,000. But when *The Wizard of Oz,* the famous story by L. Frank Baum, was being filmed by MGM in 1938, no expense was to be spared to create a frighteningly realistic tornado. The twister was, after all, one of the most vivid and important characters in the movie.

MGM's original budget for the tornado was $8,000, according to a fascinating account in a book called *The Making of the Wizard of Oz,* by Aljean Harmetz. Special-effects coordinator Arnold Gillespie's first attempt was a dud: a 35-foot-tall rubber cone, which just hung there flabbily, stiff and inert. Anyone who has ever seen a tornado, whether in real life or on film or video, knows that it is a living thing, almost continuously moving and changing. Gillespie ripped down the lifeless rubber cone and began to search for something that would better suggest a tornado's shape-shifting flexibility. Himself a pilot, he remembered that airport wind socks were continuously in motion and vaguely suggested the shape of a tornado.

That's when he hired Bethlehem Steel to build a mobile, 35-foot-high steel crane or gantry, which could move around a soundstage. Then he had an enormous, tapered muslin sock created and suspended from the top of the gantry. A rod attached to the bottom of the sock (which disappeared into a slot in the stage floor) could be used to pull the bottom of the "tornado" in one direction, while the gantry pulled it in another direction, resulting in the appearance of motion.

To create the illusion that the tornado was not a solid object (like rubber or even muslin), but instead a swirling vortex made of vapor, dust, and debris, Gillespie got the idea of spraying a powdery brown dust called fuller's earth up inside the muslin sock. Compressed air was then blasted up inside the sock through hoses attached to its base so that some of the dust passed through the porous muslin, giving its edges the hazy outlines of a real tornado.

To increase the blurriness and confusion of the storm scenes in which Auntie Em, Uncle Henry, and the hired men Hunk, Zeke, and Hickory race for the shelter of the storm cellar, two panels of glass covered with gray cotton balls to look like angry storm clouds were set up four or five feet away from the camera and moved in opposite directions. At the same time, clouds of yellowish black smoke (made from sulfur and carbon) were blown down from catwalks above the soundstage, which helped to hide the gantry from which the sock was suspended. The result was an amazingly realistic dark, sinuous funnel snaking out of great clouds of dust and then disappearing back into them.

(According to Harmetz's account, none of the stagehands used respirators during the long days of filming when these scenes, thick with clouds of dangerous dust and smoke, were created. Some of the workers became ill and choked up discolored mucus for days afterward.)

MGM sent out a press release saying that, to re-create the sound of a cyclone, the studio had hired an eminent mathematician named O. O. Ceccarini, whose "delicate mathematical calculations" to compute the sound's correct pitch, volume, and frequency ran to more than 200 pages. But though this was probably little more than Hollywood hype, what's remarkable is that even today, more than 70 years after *The Wizard of Oz* was released, this famous, three-minute scene is so realistic that it still evokes the primal fear response of a real tornado. (The short tornado sequence can be viewed on YouTube by searching under "Wizard of Oz tornado.")

One of the most lasting legacies of *The Wizard of Oz,* in fact, was the fear engendered by that scene.

"I was horrified by that tornado scene in *Wizard of Oz,* as so many generations of children were," says Jennifer Bankier, now a law professor in Halifax, Nova Scotia. "I had tornado nightmares all through my childhood."

In one particularly frightening dream, she would hear a deep voice outside her window saying, "I am the voice of the tornado," and then find herself descending "endless stairs to a place with no windows."

The daughter of two professors, she used to read about things that frightened her, and this seemed to soothe her dismay. When she discovered that people actually went out chasing tornadoes deliberately, she started visiting Stormtrack, the main storm-chasing website, and reading the chatter of chasers. Finally she decided to sign up for a "tornado tour" into the western United States—the trip wound up primarily in Kansas and the Dakotas—to confront her fears directly.

The first tornado Bankier saw was "gorgeous—twice as high as a thunderstorm, with the anvil all beautifully back-lit." This encounter was alarmingly close—"we didn't 'punch the core,' the core punched us"—but she loved the whole experience. She loved the camaraderie of the other chasers and the fact that, in the American West, the grass was always blowing. "I was dealing with all these small-pea problems at home, but for a week I was happy almost all the time." She has returned to chase again and again, all across the West, and her tornado terrors have diminished with time.

David Wolfson, a retired traffic planner, has gone out several times with veteran chaser Dean Cosgrove, who does individual tours acting as a paid guide. His fees come to about $300 a day. (Most others, however, go out on commercial tours in minivans with small groups, for a week or ten days.) Wolfson says he found the whole experience thrilling, especially the element of dramatic, rapid change.

"At two in the afternoon you're underneath plain blue skies, and two hours later, there's a 60,000-foot storm covering two counties. It's like watching the atmosphere breathe," Wolfson says. Once a supercell is created, he says, it's "like a supertanker—there is this enormous mass and momentum, with power in the rear; once it forms, it's so huge it's very slow to turn. It's just beautiful to watch."

There's something about witnessing these atmospheric extravaganzas that puts human life in its proper place, he adds.

"These storms were doing this 10,000 years before we got here, and they're going to be doing it 10,000 years after we're gone. We're just spectators."

Although David Wolfson's sense of reverent awe in the face of these magnificent storms may not produce any hard scientific data, it is, after all, an abject sense of awe that leads to scientific inquiry in the first place. Tim Samaras became a premier tornado scientist because he was first a ten-year-old boy who was enraptured by tornadoes. Cathy Finley was a frightened, fascinated farm girl before she became a professional severe-storm meteorologist. Others who have gone out on commercial tours were not scientists when they started, but they are now. Like Angie Norris, a native of East Tennessee who had no scientific training at all when she went out on her first storm-chasing trip and had a "eureka moment, and I just wanted to see more." That eureka moment led her to a quest for deeper understanding of tornadoes, with the hope of perhaps adding something to the slender body of knowledge that exists about them. She eventually moved to Norman, Oklahoma—Tornado Capital of the World—to enter a four-year meteorology program at the National Severe Storms Laboratory. Many middle-aged women look forward to grandchildren; Angie Norris is looking forward to doing tornado damage assessments for the NSSL.

Tim Samaras does not operate commercial storm-chasing tours, but he has several friends who do, and he is supportive of what they contribute to storm chasing.

"Considering how difficult it is to get close to a tornado, and how dangerous, probably the best way to go for somebody who wanted to experience storm chasing for the first time is to go out with one of these commercial groups," Tim says. "Some of these companies are operated by credible and distinguished people (and some, of course, are not). The three groups I would recommend, operated by good friends of mine, are Silver Lining Tours, run by Roger Hill; Tempest Tours, operated by Bill Reid; and Cloud 9 Tours, run by Charles Edwards."

Storm chasing has grown so popular that a storm-chasing convention in Denver in the spring of 2008 was attended by about 200 people. This annual event was born years ago as an informal get-together in Tim Samaras's basement. Its purpose is to help create a "storm-chasing community" to share experiences and information; provide a kind of continuing-education program for chasers who want to learn more; and, of course, to have fun. At the conference, there is clearly a sense of warm camaraderie among people who share a passion that they may not share even with their own spouse. Many had come to this conference from great distances to see friends they met while chasing.

Like Jacquie Kukuk, who lives in Yuma, Arizona, "where there is no weather," and who went out on a tour for her 50th wedding anniversary. She enjoyed it so much that now she goes out every spring, spending ten or twelve days hunting twisters with a group of others who share her fervor. The trips generally cost her about $3,000 each.

The Denver conference featured a half dozen distinguished speakers, such as National Weather Service meteorologist David Floyd, who gave a talk concerning guidance for storm spotters—the people who relay their severe-storm and tornado sightings through the Skywarn Network to the NWS, and thus help communities issue warnings in time to save people from tornadoes.

There was also a presentation by the "odd couple" of storm chasing, the prominent meteorologist Josh Wurman, who has partnered with a young adventurer named Sean Casey for a Discovery Channel series about storm chasers. Wurman and Casey represent the split personality of storm-chasing—one the one hand, Wurman's deeply serious search for scientific understanding of tornadoes, and on the other, Casey's deep human yearning (or is it just a male yearning?) to get very, very close to something that could kill you.

"We do have a suspicion that a tornado could loft the TIV," Casey confides to the conference goers, explaining that he no longer takes the kinds of chances he once did. "I've got two small children now, so my perspective has changed."

Wurman was the developer of the truck-mounted Doppler radar dishes known as Doppler on Wheels (DOW), used to extract high-resolution images of the tornadic vortex at close range, a genuine breakthrough in viewing not just the funnel's exterior contours but also its hidden architecture and wind velocity. (Wurman and Tim Samaras have collaborated on several scientific papers together.) Sean Casey, by contrast, is essentially a cowboy, attempting to drive his Tornado Intercept Vehicle (TIV)—a massive, eight-wheeled heavy-duty truck wrapped in two layers of steel and Kevlar mesh—directly into a tornado. The goal, he tells the conference, is to capture "the awesome spectacle of a tornado in the largest format available"—that is, a large format film. But so far, he has not succeeded.

"Look," Tim says, "what Sean is trying to do is very, very difficult. And, I don't know, maybe he has a few screws loose. But I respect him for trying."

But though Casey may not have succeeded in penetrating a tornado (at least not yet), his and Wurman's pursuit of one has made for a fascinating cable TV show, which has drawn about a million viewers a week, according to Casey.

Originally the show's producers wanted to pair the two of them with a gorgeous, doe-eyed starlet. "I think they basically wanted hunks and babes hunting and loving their way across the West," Wurman says. In the end, the producers settled on having them accompanied by a female news reporter with a science background. Still, the production crew for the television show was like a traveling circus, which required 20 hotel rooms each night, with TV people outnumbering scientists by more than three to one. Tim's TWISTEX team, chasing tornadoes across the West, often crosses paths with Wurman and Casey, whom they can see coming from miles away, with their huge truck-mounted radar dishes, the tank-like TIV, and the cavalcade of cars streaming across the countryside.

The last night of the conference is "video night," when everybody gets to showcase their best footage of the season. Many of the videos are accompanied by musical sound tracks, ranging from eerie, space-alien techno music to lush symphonic scores, but somehow none of them seem to quite capture the unearthliness, grandeur, naked fear, and heartbreaking destruction of an actual tornado touchdown.

Many of the attendees seem to go storm chasing as their primary, and sometimes only, pastime. Some have been on as many as 20 trips. One Bucks County, Pennsylvania, dentist, Thomas Howley, says he sometimes goes on two trips a season. When asked what they have given up to go chasing, various people mentioned—cheerfully and without regret—trips to Berlin and Madrid and a cruise to the eastern Mediterranean.

Many of the storm chasers at this conference had demanding, high-skill jobs—they included software designers, an IT specialist, a geologist, lawyers, doctors, a theologian. Though they were predominantly male, there were many women as well, hailing from all parts of the country and overseas. Timothy Bond, a British graduate student studying at Imperial College London, said he came 6,000 miles every summer to see twisters and to augment his understanding of the atmosphere. Like a fair number of other chasers, he had a scientific background, working on high-resolution models of ocean currents as part of his graduate studies. "The oceans and the weather are inseparable," he said.

For others, the attraction is more social than scientific. "There's just a kinship that grows up when you're with a group of people who share a passion," one said. "Sometimes people in the group understand your passion better than your own spouse."

Then there's the "tootling around the countryside," Bankier says, "and the little podunk towns that you kind of fall in love with, and all the while there's the possibility of seeing 'the big one.'"

Several of them jokingly talked about their storm chasing in the language of addiction—they called themselves addicts or said that they were hooked. One referred to "CWS"—"chase withdrawal syndrome," which sets in after the season is over and one faces the inevitable return to humdrum life. But if it's an addiction, it is a rewarding kind.

Of course, not everybody who flocks to the Great Plains to bear witness to storms is motivated by scientific curiosity or a respectful and reverent awe. At some indefinable point, a line gets crossed and storm chasing begins to turn into a kind of "disaster tourism," rubbernecking over the fearful destruction of tornadoes and the sad rubble of people's lives. And there are some chasers who disregard not just the sorrow but also the safety of people in small prairie towns.

"There are some chasers who give everybody else a black eye, stopping at nothing to get to that storm, having little regard for speed limits in small towns," Tim says. "The storm-chasing community as a whole frowns on that, because we want to have the freedom to chase without being harassed by law enforcement. After all, we're out there gathering data that will hopefully make them all safer.

"If you get one or two chasers who go flying through a little town, then sometimes law enforcement in that town will develop a sort of vengeful attitude toward all chasers, and they start to pull anybody over for anything. Then you get this antagonism between chasers and the local police, the local people. We really try to keep this from happening. We really try to be respectful of laws and of people. Some law enforcement people actually really enjoy storms—they'll pull up behind us when we're stopped and ask questions, and we try to be extremely helpful."

To be fair, the number of storm chasers who are out there simply for the cheap thrills is probably quite small. "If people get into storm chasing just for thrill-seeking, there's bound to be a lot of disappointment, because 99 percent of this is pretty boring," says longtime storm chaser Dave Hoadley. "It involves a huge amount of driving, and most of the time you don't see tornadoes. But there are some people on this Earth who just love storms, just love the beauty of the sky, and that's what makes a die-hard chaser. I like the rain, hail, sunsets, all of it. If you're just going to see tornadoes, the rewards are few and far between."

The very first bona fide storm chaser is generally considered to be Roger Jensen, a big, plainspoken Swedish farmer who started chasing and photographing storms in the summer of 1953, when he was 20. From his family's farm outside Fargo, North Dakota, he would range the countryside in the family DeSoto, a car he called "nothing real fancy," searching for big weather. Why? "Gosh, it's for the awe at what you are seeing," he told an interviewer. "I was born loving storms. I became aware of this by the time I was in the third or fourth grade. I realized right away that I was different in that way, and it's been my strongest interest all my life."

Although Roger Jensen is esteemed as one of the fathers of modern storm chasing, in fact he often did not chase storms at all. He just stayed home on the farm, or near it, peaceably observing the atmospheric spectacles as they unfolded overhead. Unlike many modern chasers, he was not obsessed with chalking up an awesome tally of tornado sightings, either. He was as interested in thunderstorms as in tornadoes or any other kind of severe weather. The important thing, he said, was to bring to all of it a sense of "awe and respect."

"He was as simple as simple gets," veteran storm chaser Tim Marshall recalls. "He would just sit out in a field and take pictures, very sparingly, sometimes four or five shots a day."

Jensen started out using Kodak box cameras but graduated to more sophisticated gear like a 35-millimeter single-lens reflex Miranda with wide-angle and telephoto lenses. He got a polarized filter to bring out the crisp, cauliflower appearance of powerful thunderstorm updrafts. Over the years he shot thousands of images of storms, tornadoes, and clouds. A particularly incredible image of golden mammatus clouds (pendulous, breastlike cloud features that often presage a big storm or sometimes a tornado) graced the cover of *Weatherwise* magazine, and other of Jensen's images were included in the international cloud atlas *Clouds of the World.* He also took pictures of the biggest (as big as softballs) hailstones ever documented in the state of Minnesota and got close enough to lightning strikes to smell the ozone after a concussion blast.

GLOSSARY

Mammatus clouds Rounded, smooth protrusions hanging from the underside of the anvil of a thunderstorm; though they accompany severe weather, they do not produce it.

He actually witnessed the great Fargo tornadoes of June 20, 1957, later made famous by Ted Fujita. One of the tornadoes touched down about 30 miles west of the Jensen homestead, though he was too busy on the farm that day to break free and go chase it. When asked later by an interviewer if he'd ever corresponded with Fujita, Jensen said: "Oh yeah. They thought I was his brother there for years. He had three or four different secretaries that knew me. I have his research paper number 42 on the Fargo tornadoes. I thought he done a really good job on that."

One of the most intense storms Jensen ever experienced, a 1967 thunderstorm near Minneapolis, resulted in some of his most amazing photographic images. After seeing his pictures, one observer wrote, "I saw one print of this storm at maximum tilt, as it advanced darkly with a solid wall of fractus 'teeth' beneath a mid-level wind-sheared shelf, that hung from above like the dark brow of an angry giant. It was one of those end-of-the-world scenes that makes one catch his breath and look for cover."

Jensen would never have used fancy language like that. But though he was a simple man who "never did go to any meteorology schools," he became knowledgeable enough that he was able to converse with any weatherman, and he corresponded with renowned meteorologists like Alan Pearson, David Ludlam, and Bernard Vonnegut (brother of the novelist).

In his declining years, Jensen had a long run of bad luck. The farm had to be sold. He moved to a small apartment with his mother. He got a dismal job at a turkey processing plant. Then his mother died. He developed diabetes, lost a leg, and wound up confined to a nursing home. But even near the end, he would hobble out to a favorite spot in a field a few hundred yards from the nursing home and sit for hours, camera in his lap, scanning the skies for storms.

When Roger Jensen died in 2001, a former Environmental Protection Agency official and artist named David Hoadley took over the unofficial title of oldest living storm chaser and photographer. Hoadley, now 70, has been chasing storms for 52 years, having gotten interested during the summer of 1956 when a huge thunderstorm rampaged through his hometown of Bismarck, North Dakota. He was 17 years old at the time, and the whole day after that storm he drove around town in the family car, filming the damage with an eight-millimeter movie camera and just marveling at it all. He became fascinated with the power and magnificence of the great prairie storms, and after that he began spending every spring chasing and photographing thunderstorms across the Dakotas.

In his quest to lay eyes on "the big one"—a real tornado—Hoadley began developing a kind of seat-of-the-pants method of trying to predict where a tornado might occur far enough in advance to get there in time to see it. This was long before computers and satellites, of course, so the only detailed and frequently updated data were hourly station reports received by teletype or "wet fax" at the local weather bureau offices. Hoadley—a gawky, enthusiastic teenager, hungry for knowledge—started going down to the local weather bureau in Bismarck, where the station meteorologists were happy to let him read the teletypes over their shoulders or look at the incoming faxes of weather maps, which were still damp when they came out of the machine and had to be hung up to dry before they could be studied properly. Hoadley made sure to time his visits for the morning, when the meteorologists were generally not too busy (storms almost always develop later in the day).

Hoadley's big day finally come on May 25, 1965. He was approaching Dodge City, Kansas, in the early afternoon when local radio began interrupting regular broadcasts with tornado warnings. "I was ecstatic," he recalls. Huge cumulonimbus cloud towers began piling up in the sky, as well as weird banks of mammatus, looking like the underside of a herd of milk cows. In the distant southwest, he could see lightning flashes. Thunder rumbled. He parked his car with a few other curious onlookers, including a policeman who was calling in to his station on a two-way radio as he watched the storm unfold. Hoadley could hear the mournful wail of tornado sirens from the surrounding towns.

Then, about three miles to the north, "a symmetrical cone funnel took shape, graceful and silent, with slowly rippling waves moving up and down the sides." He tried to take a picture with his old square-format Mamiyaflex camera, but he was shaking like a leaf. He put the camera on the hood of his car and took the only decent picture he got all day.

As the tornado moved away, he jumped back into the car and charged north, following it along with throngs of locals, some of them with "bibs still under their chins and fresh from dinner," who were taking their whole families out to see the storm. That day, he recalled later, was one "I'll always remember—the police radio; the patrolman describing some unreal, hypnotic event; the ghostly sirens moaning in the wind; and that classic cone tornado, so close and deceptively graceful as it descended to create a moment of history in the life of a small town."

Over the ensuing decades, Hoadley was out chasing every season (except for 1966, when he was on his honeymoon). His new bride, Nancy, went out chasing with him in 1967 but "quickly tired of the long, hot dusty miles. She claims we went through the same small Kansas town three times in one day!" But like many other chasers whose spouses or significant others do not share their passion, Hoadley has worked out a satisfactory marital agreement with his wife: He goes chasing; she stays home; everybody's happy. (Chaser Tony Laubach actually has a written agreement with his longtime girlfriend that his chasing will not become a bone of contention in their relationship.)

Hoadley eventually became a kind of elder statesman of storm chasing. In 1977, he founded *Stormtrack,* a bimonthly newsletter for chasers with the avowed purpose of helping to create a genuine community of chasers, for sharing experiences, exchanging views, providing tips on camera techniques, and creating a source through which commercial interests (if there were any) could buy slides or prints of severe storms and tornadoes, perhaps to learn more about them for education and safety. The newsletter was illustrated with Hoadley's deft sketches and "funnel funnies" (he studied briefly at the Corcoran Gallery of Art, near his home outside Washington, D.C.). *Stormtrack,* originally printed on paper, was reborn in electronic form on the Web and is now the premier online chaser's site *(stormtrack.org).*

"Dave Hoadley is highly respected in the storm-chasing community," Tim says. "He's a real gentleman, very thoughtful, very soft-spoken, who has been around since almost the beginning of storm chasing. He's eager to help younger people, and to share what he's learned."

Over the years Hoadley perfected his tornado-forecasting method using surface weather data. His technique, he admits, is so convoluted and idiosyncratic, filled with "obtuse terminology" based on his own observations, that probably no one else would ever use it. His method makes use of science, intuition, and experience, he says, including the discovery of visual patterns. "After thousands of hours studying hundreds of maps, I have identified certain morning patterns that are likely to produce afternoon tornadoes," he explains.

The main thing is that it works. And, in fact, Tim was impressed enough with Hoadley's technique that he invited him to make a presentation about it at the storm-chasers conference in Denver. When he carefully studied his tornado forecast record over a three-year period recently, Hoadley found his method to be 74 percent accurate in finding a "box" or forecast area where at least one funnel would form, with at least nine hours of lead time in order to get there in time to see it. Of course, he admits, he sometimes arrives at the predicted location and then picks the wrong storm to follow. But, he says, he sees roughly 4 or 5 tornadoes a year, for a grand lifetime total of 193. That's better than almost anyone else out there, including those whose vehicles are bristling with high-tech storm-tracking equipment.

"Dave 'came of age' before a lot of the new technology like Doppler radar became available, and he feels comfortable with his method," Tim says. "But I'm somebody who grew up loving electronics and computers—I just feel comfortable with it. I'm not knocking Dave's way; I guess there's just more than one way to skin a cat."

David Hoadley was also one of the first and most thoughtful chasers to attempt to answer the question: Why do storm chasers do what they do?

One of the biggest reasons is "the sheer, raw experience of confronting an elemental force of nature—uncontrolled and unpredictable," Hoadley wrote in a beautiful short essay in an early issue of *Stormtrack.* "Few life experiences can compare with the anticipation of a chaser while standing in the path of a big storm, in the gusty inflow of warm, moist gulf wind—sweeping up into a lowering, darkening cloud base, grumbling with thunder as a great engine begins to turn."

Another reason is the daunting intellectual challenge of trying to forecast accurately and consistently where a tornado will occur—an enormously difficult task.

A third reason, he writes, is "the sense of participation in a great event that comes with knowledge of the dynamics and structure of those storms. Knowing the turbulent mosaic of wind streams that weave over, around and through the towering thunderheads—and understanding their sources in the great rivers of air that sweep the continent—makes the observer almost become a part of that which he observes; as if—by force of will—he could detach himself from earth and ride the wind up into the storm's core."

It is an uplifting mystical allure that has led a small group of people to build their whole lives around the sheer, crazy love of storm chasing. (Tim Samaras is in another category, perhaps: those who have been transported by love of chasing, but have taken their love one step further, to a search for first causes, explanations and new discoveries.)

Take Verne Carlson, for instance. Verne hangs on to his job as a computer analyst at a health care company in Golden, Colorado, for one main reason: It gives him six weeks of vacation time to go storm chasing. He and his two boys, 21-year-old Michael and 19-year-old Eric, go out every season in their old Subaru crammed with wireless Internet hookups, laptops, video cameras, and several toy airplanes, whose wings are so broad they have to be detached to fit into the car. Verne's dream is to use these hobby aircraft, with remotely operable video cameras aboard, to do damage assessments after tornado touchdowns but also perhaps to fly right into or at least around a tornado funnel. (The Carlson family exploits are chronicled at *stormchaserco.blogspot.com.*)

Tim, with his Nerf sondes and UFO probes, is hardly someone to be skeptical of another ambitious tinkerer, and he's not: He admires Verne's attempts get data with toy planes, and thinks they're promising enough to include them in the daily TWISTEX e-mail reports he sends from the field.

Carlson calls these aerial surveillance missions the Wicked Witch Project. Another project that tips a hat to *The Wizard of Oz* is the Flying Monkey Video Downlink, which would use a $4,000 model helicopter instead of a $500 RC glider to fly over tornado damage paths. The little chopper would be able to hover when necessary. The new system, using a much improved high-definition video camera, would allow an operator on the ground to point and tilt the camera by means of remote control. Verne, Mike, and Eric Carlson are the only chasers trying to do this, at least so far, but if they're successful they are sure to spawn imitators.

Though the Flying Monkey project is still relatively elementary, it naturally suggests some perhaps soon-to-be-invented technology: the means by which some superhardened drone aircraft, equipped with video cameras, could fly into a tornado while a person on the ground wearing video goggles and earphones could directly experience "getting inside" the world's most violent storms firsthand. (So far, Tim's video-laden HITPR probe is probably the closest thing to such a device. In fact, the Field Museum in Chicago has created an exhibit using Tim's video and soundtrack to simulate the experience of "being inside" a tornado.)

Verne Carlson, who has seen 97 tornadoes over the past 20 years, got interested in storm chasing almost by accident. He and a couple of friends were returning from a caving trip to the Black Hills of South Dakota when they topped the crest of a hill and saw a police car, lights flashing, pulled off the road in front of an immense, black, oncoming storm. They stopped the car and watched as "all these tumbleweeds in a field started doing this crazy dance, and then, maybe 30 seconds later, these long, black fingers formed and reached down, and then they turned into a long needle, and then that turned into a large cone tornado."

After that experience, Verne Carlson was hooked.

"Tornadoes look like something that shouldn't even be on this planet," he says.

When he discovered that there were people who actually went out hunting tornadoes, in order to increase their odds of seeing one, he decided to join the club. Now, he says, his odds of seeing a tornado are about one out of every ten trips, less than Hoadley but probably fairly typical of chasers' experience in general.

Though Carlson's wife has no interest in chasing tornadoes, he says she is completely supportive of the passion he and the boys share. Recently he bought a cheap house in Amarillo, Texas, about a day's drive from his home in Colorado, solely to use as a storm-chasing pit stop. Now he's looking at buying a little house in Russell, Kansas. Other people buy real estate for the view or the school district; Verne buys it for tornado probability.

"These little towns are amazing," he says. "You can buy a decent two-bedroom house with a garage and a small yard in town for almost nothing. I'll just put antifreeze in the lines during the winter and mothball the place until storm-chasing season in the spring."

Anything to get out there and see another twister.

Quite a few chase groups are out-of-the-classroom meteorology workshops, affiliated with universities like Texas Tech or the University of Oklahoma. Kevin Myatt, who writes a weather column for the *Roanoke Times,* in Roanoke, Virginia, was one of the co-leaders of a chase group of ten students from Virginia Tech and other schools who converged on WaKeeney, Kansas, during spring 2008. The group was divided into two minivans, with Myatt driving one and the other driven by co-leader Dave Carroll, a meteorology teacher with 20 years of storm-chasing experience.

Normally, Myatt says, he tells the students that they are supposed to be "observers, not participants" in the storms they chase. But on the afternoon of May 22, near WaKeeney, the group came entirely too close to becoming the story. That's when they ran into the same astonishing series of supercell thunderstorms that Tim Samaras and the TWISTEX team tangled with all that day.

"It was around 7:30 and starting to get dark," Myatt recalls. "We thought we were at the end of an incredible chasing day, with our running tornado count at seven, coming out of three different supercells. There was another big storm forming south of WaKeeney, and our first concern was to be somewhere outside that storm system.

"But our attempt to stay out of the way pretty quickly turned into a full-fledged chase (either that or a full-fledged flight). We pulled off on a gravel road and the thunder and lightning were just continuous. It was a beast of a storm. You could hear the tornado sirens going off. We could see a wall cloud forming that was suggestive of a cone beginning to take shape, so we continued south toward the storm. I was in the lead van and Dave was in the other van, watching the radar. He said to turn right onto a gravel road south of WaKeeney, off 283, and then we could see this lowering cone with incredible yellow gold backlighting. It was one of the most surreal things I've ever seen, as the cone lowered and lowered all the way down to the ground. It was gorgeous. We could not have been more than a mile to a half mile away.

"Unfortunately we had seen so much action that day my video camera was out of juice, so I grabbed my digital camera. Then all of a sudden we started getting wrapped up in rain and you couldn't see the cone very well anymore. To the south of us, a wall of rain was starting to move in.

"'We gotta get outta here, Dave,' I told him over the radio. 'We need to go or we're gonna get hooked' [trapped in the tornadic core]. There were so many chasers on this muddy gravel road you couldn't really turn around, so we just backed down the road and got back on 283, south of WaKeeney. Then the hook started to arrive. There was hail about the size of marbles, and rear-flank downdrafts, probably 80-mile-an-hour winds. We really, really didn't want to be in that situation. I was actually more fearful of baseball-sized hail than catching the fringe of the storm, but now we were pretty much caught in both.

"In back of the van, the girls had their coats over their heads, and the guys had their heads between their knees. It was a pretty tense five minutes. I prayed a silent prayer for the wind to shift from the southwest. When it shifted, I knew it was following the storm and moving away from us. Then I heard Dave crackling over the radio that the other van had gotten out of it. We drove a couple of miles east and had a reunion and a debriefing, with lots of hugs and happiness. But if you chase storms, sooner or later you're going to get into a situation that's a little dicey like that."

"A little dicey like that": Those are the moments that many storm chasers live for.

Tim Samaras has certainly had his fair share of dicey moments. And though he has a serious purpose in being out there, there are times when he just grins at the pure excitement of it.

Veteran chaser Tim Marshall, who has been at it since 1978 and has seen more than 200 tornadoes, is a meteorologist, a civil engineer, and a man who loves severe storms more than almost anything else in this world. Not long after an F4 touched down four miles from his home in Oak Lawn, Illinois, when he was 11 years old, "I knew I would be studying tornadoes for the rest of my life," he says. He joined the Weather Club at his school and later went on to graduate studies in atmospheric sciences and engineering at Texas Tech University. The dual major was to prepare himself for the work he's been doing for the past 25 years: conducting tornado damage assessments for a wide range of clients, from insurance companies to the National Weather Service. What he has seen is enough to sober up even the most storm-drunk chaser.

Take the frightful aftermath he found in Jarrell, Texas, after an F5 virtually obliterated the town on May 27, 1997. The storm was moving at only nine miles an hour, which meant that the tornado could linger over any given point for as long as three minutes, essentially grinding anything inside the vortex to bits. Afterward, Marshall found a car that had passed through this tornadic buzz saw, which had shredded the car so completely that nothing was left but the engine block.

Human bodies had also been shredded in the same way, in one case leaving nothing but a torso with one arm, a hand, and the wedding ring that enabled investigators to identify the corpse. The tornado's ferocity was surreal, ripping the grass right out of the ground, stripping asphalt from highways, debarking trees, dismembering hundreds of cattle and completely sweeping away houses, even peeling back the linoleum on wrecked, roofless kitchen floors. Twenty-seven people were killed.

"That's about as bad as a tornado gets," he says. Did this experience change his reason for storm chasing? No.

But Tim Marshall is not a sadist or an ambulance chaser. He does not revel in schadenfreude. The Jarrell tornado was so extraordinarily rare it almost qualifies as a freak event, he points out. Most tornadoes are actually relatively weak. And like Roger Jensen, it is not so much their destructiveness as their astonishing beauty that captivates him: "There are some people on this Earth who just love storms and the beauty of the sky—the rain, the hail, the supercells, the sunsets," he says. "The sky is so huge, you are immersed in something much bigger than you are."

That about sums up his motivation for going out there year after year to pursue, document, and understand big weather. And unlike 30 years ago, when he started out, he is no longer nearly alone. In the early days, as storm chasing caught on among a small, self-selected group of aficionados, perhaps as many as half of these people were either working meteorologists or people with a meteorology background, David Hoadley says. Then, especially after the release of the 1996 movie *Twister,* there was an explosion of interest in storm chasing, with thousands of people participating in commercial tours, and millions more watching storm-chasing programs and videos on cable TV and on the Web. (At least one tornado video posted on YouTube has had more than three million hits.)

Storm chasing has come of age. But is that a good thing or a bad thing?

In a 1977 issue of *Stormtrack,* long before chasing became popular, Hoadley observed that "one of my long-standing concerns has been that stormchasers will eventually draw too much publicity, and chasing will become another mass cult of the leisure class."

Well, Hoadley says today, it's happened. And to the extent that this has been a detriment to science, making it more difficult and dangerous for bona fide researchers like Tim Samaras to get in close to tornadoes, well, that's a problem.

In addition, notes Hoadley, "Over the decades, as this hobby has developed, with tour groups and so on, it has turned into a bit of a circus out there. And with crowded, muddy roads where it's possible to get trapped in a storm, there's good reason to be concerned."

The biggest danger, says Marshall, is the driving. Chasers go flying down muddy, unfamiliar roads in pelting rain or hail, or drive too fast through small towns, so intent on the pursuit of the twister that they're not paying attention to other vehicles on the road. Lightning is another potentially fatal hazard. "I've had a couple of close calls, so now I only film from inside the car," he says. But many people don't. Crowds of storm-watchers will stand in the road or on the roofs of cars to film a big storm spectacle, practically inviting disaster.

In fact, Marshall says, some tragic event is probably inevitable, with so many people getting closer and closer to something so dangerous.

"On the other hand, you can't prevent people from enjoying the wonders of nature," Hoadley says. "I think on balance, the more people know about these storms, the better. People go home and tell their friends, and this knowledge and awareness just ripples outward, like a stone splashing in a pond, and that's to the good."

"I like to have a pretty low-key, unjudging attitude toward the whole storm-chasing circus out there. I just assume they are mostly people with good intentions, inspired by the same things I'm inspired by. Just so long as I can get my work done, I'm happy to be part of it."

13: GLOBAL WARMING AND TORNADOES

THERE WERE A COUPLE OF FREAKISH THINGS ABOUT THE OUTBREAK OF TORNADOES that rampaged through Tennessee, Arkansas, and other southern states on February 5 and 6, 2008. This savage, long-lasting storm complex spawned more than 80 tornadoes, including five rare and ruinous EF4s, which killed 57 people. Taken together, the storms caused more than $500 million in damage, making the tornado outbreak the fourth costliest in U.S. history.

One of the most peculiar things about the so-called Super Tuesday Outbreak (so named by the media because of the large number of presidential primaries earlier that day) was its timing. It was very nearly the dead of winter, whereas violent tornadoes are usually creatures of the warm, wet, turbulent atmosphere of spring or early summer. One researcher pointed out that only one other tornado outbreak in the past century had killed so many people so early in the season—and for that you had to go back almost 60 years, to an Arkansas tornado that killed over 50 people in 1949.

Another strange thing was that although tornadoes usually don't make contact with the ground for more than 10 or 15 minutes, many of these twisters were deadly "long-track" tornadoes that buzz sawed across the surface of the earth for enormous distances. One in Arkansas left a damage track 122 miles long, an all-time record for the state. One more oddity was that the whole South had been basking in record-breaking summerlike warmth at the time. It was 75 degrees in Little Rock, Arkansas, that winter day, and 78 in Shreveport, Louisiana.

"The weather service has told us we are going to have more and more intense storms," prominent Democratic senator and former presidential candidate John Kerry said in a TV interview, "and insurance companies are beginning to look at this issue and understand this is related . . . to the warming of the earth." Though he had given voice in public to suspicions many people harbored in private, Senator Kerry was immediately assailed by right-wing bloggers who mocked him, the "hoax" of global warming, and the suggestion that tornadoes could be linked in any way to (nonexistent) climate change. Left-wing bloggers quickly returned fire, and an angry battle ensued. The intensity of the argument showed just how provocative, emotional, and complex the whole issue is. But what was mostly left unspoken in the fight was a calm, reasonable, scientific discussion of the seemingly simple question: Will global climate change increase the frequency or intensity of tornadoes?

Part of the reason the question was so ripe for partisan sniping was that, actually, it's not a simple question at all. In fact, like almost everything else involving Earth's tumultuous atmosphere, the question is so dauntingly complex that it may be impossible to answer (at least, at the current level of scientific understanding). Meteorology, says storm-chasing meteorologist Tim Marshall, is an inexact science, filled with probabilities rather than certainties. It's not like pure mathematics, where there is a right answer, a wrong answer, and nothing in between. Which is why the dead certainty of advocates from both sides is wrong-headed.

"Whenever I give lectures, for kids or adults, somebody always asks about tornadoes and global warming," Tim Samaras says. "It's a whole lot easier to ask the question than it is to answer it."

To begin with, explains meteorologist Roger Edwards of the National Weather Service's Storm Prediction Center, it's extremely difficult to relate small-scale, daily fluctuations (the weather) to large-scale, long-term trends (the climate). There are "huge differences in size, time scale, and physics" between global average temperature shifts and tornadoes, which are "too small, too short-lived, hard to measure and count, and too dependent on day-to-day, even minute-to-minute weather conditions" to make these connections. The bottom line: "No scientific studies solidly relate global temperature trends to tornadoes."

Even the question itself shows how little is actually known about tornadoes. "We still don't understand exactly how and why tornadoes form, because it is so hard to understand what we can't measure well," Edwards explains. Even with truck-mounted mobile Doppler radar and instrument probes, "we have tremendous trouble sampling even a tiny fraction of all the tornadoes that do occur."

Which is why Tim Samaras's HITPR probes are so valuable. His quest to deploy measuring instruments directly in front of oncoming tornadoes has produced a steady stream of data that have helped researchers understand how these winds are formed, how they feed and grow, and how they strengthen or languish. Still, he says, "we are working in such a fine scale, a 'meso' [medium-size] scale, just focusing on individual thunderstorms and tornadoes, and not the climate as a whole, that I think our work will contribute to the understanding of tornadoes and climate change—if it exists—only in a very indirect way."

Even so, many of the "simplest" questions remain baffling. "We still don't understand . . . why some thunderstorms produce tornadoes and others don't," Edwards says. In fact, severe-weather meteorologists don't even completely agree on how to define a tornado at all.

Another problem with trying to answer this question is the unreliability of tornado counts. It's simply not known with any degree of certainty how many tornadoes have occurred each year in the United States or elsewhere in the world for any extended period of time. It's impossible to know how many tornadoes have occurred in the vast emptiness of the

High Plains without being seen or reported, or how many dust storms or ordinary thunderstorms have been misreported as tornadoes. (The best estimates are that about 1,300 tornadoes occur, on average, each year in the United States, though this number varies considerably.)

"The U.S. has the most detailed tornado reporting system in existence," Edwards observes, "and several recently published scientific studies have shown that even our roughly half century of 'good' tornado data still can be very deeply flawed. . . . Elsewhere in the world, tornado records are so spotty and inconsistent that they're next to useless."

If it's not known how many tornadoes occurred in the past—the "normal" baseline—how can it be known if their number is increasing?

Nevertheless, tornado statistics (especially fatality statistics) have often been trotted out in the heat of the debate. Global warming naysayers have pointed out that the top ten deadliest years for tornadoes all occurred before 1954, and that the deadliest tornado decade on record was the 1930s. In fact, the number of tornado fatalities has declined steadily every decade since the 1930s, except for a slight uptick during the 1970s. On the other hand, 2008 turned out to be the deadliest tornado season since 1953, with 125 fatalities. In the month of May alone—when Tim and the TWISTEX team encountered a relentless onslaught of tornadoes, including the massive EF4 near Quinter, Kansas—43 people were killed.

Those are only *fatality* statistics, though. It could be that far fewer people are killed by tornadoes today because weather satellites, radar, and other technologies have improved early warning systems. The numbers don't say whether tornado frequency is increasing, decreasing, or staying the same.

"I've been storm chasing for around 20 years, and it does seem to me that Tornado Alley is moving a little bit north," Tim says. "We seem to be finding fewer tornadoes down in Oklahoma and Texas, and more up in Kansas, Nebraska, and South Dakota. But that's just anecdotal evidence, not real scientific evidence. There may be a trend there, and there may not be."

In recent decades there has been an increase in the number of tornadoes reported, with an all-time high of more than 900 occurring in the year 2002, and about 1,500 in 2008. This, the National Oceanic and Atmospheric Administration (NOAA) says, is due to "increased national Doppler radar coverage, increasing population, and greater attention to tornado reporting. . . .This can create a misleading appearance of an increasing trend in actual tornado frequency."

Thousands of storm chasers swarming over the West every summer would certainly tend to increase the number of tornadoes that make it onto the books.

What happens if you just strip out all the smaller, weaker tornadoes, which might have gone unnoticed without storm chasers, and include only the most extreme and violent ones? NOAA data that include only the EF3s, EF4s, and EF5s do not show a steady increase in frequency over recent decades. In fact, the highest number of these superviolent tornadoes occurred in 1974.

What if it were possible to just forget about the unreliable tornado statistics—and even the past itself—and somehow "see into" the warmer future world that many atmospheric scientists have predicted?

In 2007, scientists at NASA's Goddard Institute for Space Studies developed a new computerized climate model that attempted to do just that. When compared with the real world, this virtual-reality computer model held up well, showing accurate variations in severe storms over land and the oceans, and simulating hot spots of storm-produced lightning in Africa and the Amazon Basin.

Then the model was applied to a hypothetical future climate with twice the current carbon dioxide levels and an average increase in temperature of five degrees Fahrenheit. The model showed an increase in severe storms and tornadoes, especially in the central and eastern U.S.

GLOSSARY

Carbon dioxide CO_2; an odorless and colorless gas and the fourth most abundant component of dry air

Updraft A small column of rising air; if the air is moist, then the moisture will condense and form a cumulus cloud.

On the other hand—and there's always an "on the other hand" when it comes to global climate change—probably the world's most authoritative source on such questions is the U.N.'s Intergovernmental Panel on Climate Change, composed of hundreds of scientists from around the world. In its most recent report, issued in 2007, the IPCC had this to say on the subject of climate change and tornadoes: "There is insufficient evidence to determine whether trends exist . . . in small-scale phenomena such as tornadoes, hail, lightning, and dust storms."

GLOSSARY

Hurricane A tropical windstorm with sustained surface winds blowing at least 74 miles an hour (64 knots), they usually occur in the Atlantic, Caribbean, or the Gulf of Mexico.

El Niño The warming of ocean currents along the coast of South America that is often associated with dramatic changes in regional weather patterns

Tropical storm A tropical cyclone with sustained surface winds blowing at least 39 to 73 miles an hour (39 to 63 knots)

But if tornadoes are too small-scale a weather event to be affected by climate change (or at least detected by scientists), what about other, larger atmospheric phenomena, like hurricanes, the planet's überstorms? Tim Flannery, the scientist-author of *The Weather Makers,* a book about climate change, points out that "over the past decade, the world has seen the most powerful El Niño ever recorded (1997-98), the most devastating hurricane in 200 years (Mitch, 1998), the hottest European summer on record (2003), the first South Atlantic hurricane ever (2002), and one of the worst storm seasons ever experienced in Florida (2004)."

Even so, it's still a controversial point as to whether a warming Earth and its oceans would result in more frequent or more violent hurricanes. A 2005 study in the journal *Nature* by Kerry Emanuel, a professor of atmospheric science at MIT and an authority on hurricanes, suggested that this may already be happening. The study showed that the duration and the strength of hurricanes have increased by about 50 percent over the past three decades, primarily because of a warming of sea-surface temperatures, the primary engine that stokes hurricanes.

GLOBAL WARMING AND TORNADOES

A 2008 report from NOAA's National Climatic Data Center that synthesized the findings of more than 100 academic papers, concluded that "we are now witnessing and will increasingly experience more extreme weather and climate events." The paper noted the "substantial" increase in the intensity of hurricanes and tropical storms since 1970, though the scientists said there is no conclusive proof that this is connected to human activity. The paper also noted that computer models have shown that by the end of the 21st century, intense storms that produce huge precipitation events—rain, snow, hail—that now occur once every 20 years, will probably take place every 5 years. Whether these violent storms will be more likely to produce tornadoes is impossible to say.

Tim Flannery points out that relatively small temperature changes in the atmosphere can have dramatic effects. Hurricanes, he says, are powered by the latent heat released when huge quantities of water vapor condenses. "It's not widely appreciated just how much extra latent heat the hot air engendered by climate change can carry. For every 18°F increase in its temperature, the amount of water vapor that the air can hold doubles; thus air at 86°F can hold four times as much 'hurricane fuel' as air at 50°F."

So it's not beyond the realm of reason that a relatively small boost in atmospheric temperature could have a dramatic effect on the likelihood of devastating storms, including tornadoes. Then again, just because something seems reasonable doesn't mean it's true.

For now, the definitive answer to the question of how climate change might affect tornadoes remains beyond reach. Says Roger Edwards, of the SPC, it involves trying to "predict future worldwide changes in something we haven't sharply defined, can't even count or measure very well, and that we often can't predict an hour from now."

So what does it all add up to? Does climate change actually affect tornadoes? "We simply don't know," Edwards says.

14: YOUR TEN-SECOND GETAWAY PLAN

MANY INDIVIDUALS HAVE PLAYED STARRING ROLES IN THE PREVIOUS CHAPTERS OF this book. In this short but very important chapter, though, it would be helpful for you to put *yourself* at the center of the story. Because public safety specialists say that one of the most important things you can do to anticipate a close encounter with a tornado is to *have a plan in advance*. That means imagining yourself living through such a terrifying scenario.

This is not intended to make you feel unnecessarily alarmed. Your chances of being injured or killed by a tornado are actually very slim. According to tornado researcher Tom Grazulis, "less than one percent of the American population will ever be in the path of even the weakest tornado during their lives." And your chance of tangling with an almost inconceivably violent EF4 or EF5 is even slimmer. These ultradangerous tornadoes account for some 70 percent of all tornado fatalities; they are so savage that almost no ordinary man-made structure can withstand them. But the good news is that they account for only about one percent of all tornadoes. More than likely, if you do encounter a tornado, it will be a relatively weak and survivable one.

And your odds of being fatally injured by a tornado are even more remote. In recent years there have been only about 60 tornado fatalities in the United States per year (although in 2008, an exceptionally deadly year, there were more than 120). That puts your odds of being killed by a tornado at about 1 in 60,000, which is actually a bit more than your risk of getting struck by lightning (1 in about 84,000; the chance of your being struck in any given year is 1 in 700,000). By contrast, your chances of getting killed in a car wreck are about one in a hundred, according to the National Center for Health Statistics.

Still, it could happen. And despite all the advances in severe-weather forecasting technology, advance warning of a tornado usually does not give you much more than ten minutes' notice, and often less, which is why you need to know precisely what you would do long before an actual warning is sounded. Ask yourself: Where would I go for safety if I had ten minutes of warning? Where would I go if I had ten seconds?

It's a good idea to have a family tornado drill once a year, and have a predetermined place to meet afterward, especially if you live in a tornado-prone area, according to severe-storm meteorologist Roger Edwards of the Storm Prediction Center. What's a tornado-prone area? That's not exactly easy to say. If you live in Tornado Alley you no doubt know it; but "Tornado Alley" is a very loose and unofficial term, not defined by the National Weather Service or any other scientific body. It generally refers to the part of the country where the strongest tornadoes occur, the most often, which is a huge swath of the High Plains, from northern Texas and the Texas Panhandle north through Oklahoma, Kansas, and Nebraska and into South Dakota.

Unfortunately, living outside this area does not mean you are safe from tornadoes. In fact, if Tornado Alley includes the most *frequent* tornadoes rather than simply the strongest ones, it becomes a much larger area. In addition to the High Plains, this tornado danger zone would include parts of Mississippi, Alabama, and Georgia (sometimes called Dixie Alley); Florida; Arkansas; southern Indiana; eastern Iowa;

western Pennsylvania; and parts of southern New York, Massachusetts, and Connecticut. Actually, more people have been killed by tornadoes in Dixie Alley than in the High Plains, probably because of greater population density in the South. And by some measures, the most tornado-prone state in the Union is Florida (though Florida's tornadoes tend to be weaker and more short-lived than elsewhere). In fact, about the only part of the United States more or less impervious to tornadoes is Alaska. (Between 1970 and 1995, there was a total of one confirmed tornado in the state.)

It helps to be aware *when* tornadoes are likely to occur, as well as where. Although there isn't really a "tornado season," most tornadoes occur in spring or early summer (March, April, May, or June) on hot, humid, stormy days. And they usually occur in the afternoon and evening, between 3 p.m. and 9 p.m., although in the Southeast, morning tornadoes are almost equally common, according to the *Tornado Project Online*. In fact, "there is no time of year and no time of day that tornadoes do not occur," even in places where there is three feet of snow on the ground, severe-storm meteorologist Howard Bluestein observes in his book *Tornado Alley*.

So unfortunately, tornadoes can occur in any season, virtually anywhere in the continental United States. And all too often tornadoes occur with little or no warning; as a result, "there is no substitute for staying alert to the sky," observes Edwards.

Be alert for weather events that may foretell a tornado in the making: an oncoming storm that produces a sickly greenish or greenish black color to the sky; a large, dark, low-lying cloud, especially if it appears to be rotating; debris falling out of the sky; large hail; or a continuous roar or rumble that doesn't fade in a few seconds, like thunder. In a thunderstorm at night, you may see blue-green or white flashes on the ground (as opposed to lightning bolts aloft), meaning that ferocious winds—possibly those of a tornado—are snapping power lines. And if a big storm is rolling in, beware if leaves and other debris begin getting sucked upward. This inflow suggests a tornado could be forming.

The undeniable proof, of course, is when you actually lay eyes on a funnel cloud. But don't forget that tornadoes may sometimes be virtually transparent and therefore invisible, becoming visible only when they suck up dirt or debris, or when moisture drawn up into the funnel begins to condense into whitish vapor. They may also pass from visibility back into invisibility, but still remain extremely dangerous. Remember that if you can see the funnel and it is not moving to the left or the right, it is probably coming right at you.

Part of being prepared for a tornado encounter is figuring out in advance where you would go to get good weather information, whether it's the Weather Channel, the Internet, or local radio. You can also buy a NOAA Weather Radio for $30 to $80, on which you can play all weather, all the time, or that you can set it to be activated only when a severe-weather warning has been issued.

GLOSSARY

Tornado watch Issued by the National Weather Service when conditions are favorable for a tornado

Tornado warning Issued by the National Weather Service when a tornado is sighted by spotters or indicated on radar

Remember that there is a big difference between a tornado watch and a tornado warning. A watch means that tornadoes and other kinds of severe weather are likely in your area over the next couple of hours. A tornado warning means either that Doppler radar has shown the sort of thunderstorm circulation that could give birth to a tornado very soon or that a tornado has actually been spotted. The key thing about a tornado warning is its urgency. Once it's issued, it's time to seek safe shelter *immediately*.

Don't depend on tornado sirens. Although the National Weather Service issues tornado watches and warnings, it has no control over siren policy, which varies a lot from place to place, according to Edwards. It's up to local governments to have a community readiness system in place. But local governments may have no tornado sirens because they can't afford them or they feel tornadoes are too unlikely to be worth the expenditure, or the siren may have been activated so infrequently that the battery is dead.

Once you've heard a warning or seen the funnel, remember four simple things:

- Get as low to the ground as possible.
- Get behind as many walls as possible.
- Stay away from windows.
- Beware of flying debris, which represents the biggest danger.

"Your first reaction should be to head for shelter, not to grab your camcorder and run to the window. This may be very hard for some people, given that the most spectacular thing they will ever witness is getting closer and more impressive every second," writes Tom Grazulis in his book *The Tornado: Nature's Ultimate Windstorm.* "But the flying glass and incoming debris could make it the last thing you ever witness."

Don't bother to open the windows to "equalize pressure." This outdated advice is "absolutely useless, a precious waste of time," according to Edwards. If a tornado hits the house, it will blast the windows open anyway, if it doesn't knock down the walls. But the biggest danger is being struck by flying glass.

If you're in an ordinary frame house with a basement, the best thing to do is to run down to the basement and crawl underneath a sturdy workbench, or at least a mattress or sleeping bag, to protect you from flying debris. People have been injured or killed in basements when the tornado tore the whole house away and then dumped heavy objects on top of them. Find a place that's not directly underneath heavy objects (appliances, pianos) on the floor above. And don't bother about trying to find the southwest corner of the basement, which is sometimes said to be the safest place (the reasoning being that because tornadoes usually come from the southwest, debris will tend to fall on the northeast side). This is more useless advice, says Edwards, because tornadoes can come from any direction.

If your house has no basement, go to a small, windowless, first-floor interior room like a closet or a bathroom. Bathrooms tend to hold together because of the plumbing and all the extra framing required in construction. The bathtub and toilet are anchored to the ground; after severe tornadoes they are sometimes the only thing left of a house. Try climbing into the bathtub and covering yourself with a pillow or blanket, anything—even a metal trash can—to help protect you from debris.

If you have reason to be extremely concerned, you might consider reinforcing an interior room or building a tornado "safe house" or wind shelter in your basement. Federal Emergency Management Agency (FEMA) guidelines point out that although your house may be built to code, that still doesn't mean it can withstand extreme-wind events like tornadoes. You can build a safe room in your basement; on a concrete slab, such as a garage floor; or in an interior, windowless room on the first floor. The safe room must be designed to protect its occupants from wind and debris even if the rest of the house is destroyed, or nearly so. It must be adequately anchored to avoid being lifted up and turned over. The walls, ceiling, and door must be able to withstand penetration by debris. The connections between all parts of the room must be able to withstand tornadic winds. And its walls must be separated from the main residence, so that destruction of the house will not destroy the room. (Details can be found in the FEMA manual called *Taking Shelter From the Storm: Building a Safe Room Inside Your House,* FEMA-320, available at *fema.gov.*)

What if a tornado touches down while you're in a car? Cars are "notorious death traps," says Edwards. Most tornado fatalities occur in cars or mobile homes. Still, if the tornado is fairly distant and the roads

Destruction left in the wake of a fierce F4 category tornado includes a flattened mobile home.

are fairly free of traffic, you can likely escape by driving at right angles to the tornado's direction of movement. But FEMA advises that, if you're in a congested urban area where you might get caught in traffic, it's a bad idea to attempt to outrun a tornado. Although the average speed of a tornado is about 30 miles an hour, they can travel as fast as 70 miles an hour, and can change directions without warning. If the tornado is bearing down on you at very close quarters, abandon the car and lie facedown in a ditch or other low-lying area. You can wash off the mud later.

What about pulling off the highway and parking underneath a bridge or overpass? "Absolutely not!" says Edwards. "Stopping under a bridge is a very dangerous idea." For one thing, says Tim Samaras, "the laws of physics are working against you—you've got large volumes of air being squeezed through a narrow 'bottleneck' (the opening under the bridge), which means that the airspeed will actually be increased. That puts you in more danger than you would be in if you just lay down in a field."

For another, adds Edwards, if you climb up the embankment under the bridge, you're very exposed to flying debris. The bridge itself could collapse on top of you (it's impossible to judge the structural integrity of a bridge just by looking at it, especially in the haste and confusion of a storm). And parking along a highway in blinding rain or hail is extremely dangerous, for you and for other drivers—your vehicle could be rammed from behind, with tragic results. You're better off lying down in a low, flat location.

To get a vivid idea what it's like to be cowering under a highway overpass when a tornado passes over, try watching an infamous video on YouTube. (You can find it by searching under "overpass tornado.") This sequence was captured on videotape when a film crew got too close to a tornado near Andover, Kansas, on April 26, 1991, and sought shelter under a bridge. Though the encounter was hair-raising, with ferocious, roaring wind hungrily sucking everything out from underneath the bridge, Edwards points out that this was only wind from a surface inflow jet, a belt of intense wind being sucked up into the base of the tornado—not the tornado vortex itself. Being caught in the vortex would very likely have produced no videotape at all, only statistics.

What if you're caught in a school or a public building? FEMA recommends that you find your way to an interior room or hall on the ground floor, avoiding halls that open to the outside. Also avoid auditoriums or any other places with large free-span ceilings that could collapse. Crouch down and make yourself as small a "target" as possible, covering your head with your hands or anything else that's handy and might offer more protection.

What about mobile homes? The answer is simple: Get out! Mobile homes are extremely dangerous places to ride out a tornado, and should be avoided at all costs. Even a mobile home that's tied down is more dangerous than the nearest sturdy building or, if there isn't one available, low ground, where you can lie flat with your head covered. In fact, because scenes of ravaged trailer parks are such a common video clip in the aftermath of tornadoes, there is a persistent myth that mobile homes actually *attract* tornadoes. (It's not true.) A similar structure, similarly dangerous, is a frame house sitting on blocks rather than bolted to a foundation. Flee any of these structures and find your way to a place of safety, keeping in mind the four main things:

- Get as low to the ground as possible.
- Get behind as many walls as possible.
- Stay away from windows.
- Beware of flying debris, which represents the biggest danger.

If you heed this advice, you'll more than likely emerge from your unasked-for starring role with your life, your health, and a renewed reverence for the savage majesty of tornadoes.

15: PASSION AND PURPOSE

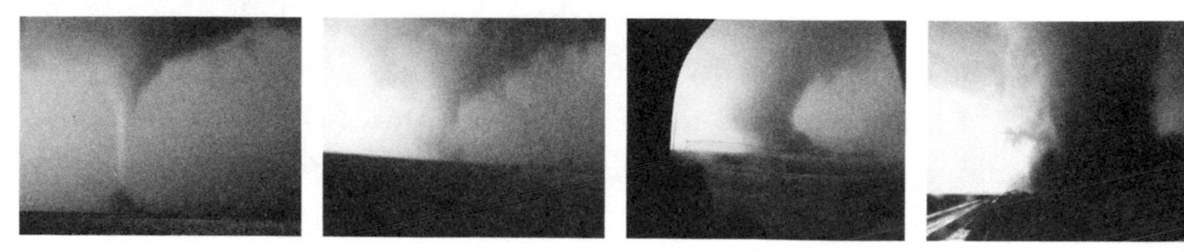

ONE HUNDRED AND TWENTY SCHOOLCHILDREN, AGES SIX TO TWELVE, ARE ASSEMBLING in a basement classroom at the Houston Museum of Natural Science, in anticipation of a presentation by a man billed as Storm Chaser Tim Samaras. Tim has been invited to Houston to do a series of shows and lectures about his work, both for adults and children, here at the museum and at a fancy banquet at the lavish River Oaks Country Club. But it's the kids' events he likes best.

As the children sort into little clots of friends and take their seats, the energy in the room is skittish, unsettled. One kid keeps dropping a pencil on the floor, as if to double-check gravity. Two boys commence firing with finger pistols. A little girl, one foot tucked up underneath her bottom, starts rhythmically kicking her other leg out in front of her.

After a brief introduction, Tim steps up to the front of the room and launches right into his story. He seems completely at ease in this setting, working to capture his youthful audience by channeling his own childhood self.

"People say to me, 'Tim, how'd you get started in all this?' and I say, 'Well, it all started when I was about eight or ten years old and I had this wild curiosity about how things work,'" he tells the kids. "My parents were always getting after me because I'd be taking various appliances up into my bedroom and taking them apart to see how they worked. I'd take the blender apart. I'd take the record player apart. The last straw was when I took the television set apart and I found all these strange tubes in there. It was *totally* fascinating!"

A murmur of laughter goes through the group. The boy stops dropping the pencil, gravity confirmed.

"I always wanted to be one of those mad scientists—you know, the guy with the gray hair that's sticking up and the wild expression on his face."

"Well, finally one day my mother knocked on my bedroom door and she said, 'Timmy, why don't you come and watch this musical on TV?'

"Aw, Mom, I'm right in the middle of this very important scientific experiment!"

But Timmy went anyway and flopped down in front of the TV. In those days, he tells the kids, "we didn't have any DVDs, no TiVo, only two channels, and most of the programs were black-and-white. Can you believe that?"

A ripple of disbelieving laughter.

"Well, this musical was about a young girl and her little black dog, out in Kansas or someplace, and I was falling asleep, when all of a sudden it came on the screen. The tornado! That tornado was the coolest thing I had ever seen. The tornado carried the girl to the land of Oz, and then the movie turned into Technicolor, and there were all these Munchkins and whatnot, and then I started falling asleep again. But that tornado is what I remember, and my passion for tornadoes and thunderstorms has remained to this day. Thunderstorms are simply astounding—better than any video game you could ever play."

Now the kids have grown almost completely still, so Tim retells, for the umpteenth time, another of the key stories of his childhood.

"My mom used to always try to get me out of my bedroom, so she enrolled me in Cub Scouts. That didn't really take, so she tried to get me to play softball with the other boys. She'd say, 'You need to get out of your bedroom, get some exericise!' So I joined a team and they put me out in the outfield. I'd just stand out there watching these beautiful thunderstorms come rolling in over the Rocky Mountains, and I'd hope it would start to rain so I could go home.

"One day I was out there watching this *beautiful* cloud formation taking shape over the mountains, just piling higher and higher and getting ready to 'fire' into a big, nasty thunderstorm. Then I looked over at the bench and everybody was jumping up and down and yelling, pointing at me. I looked behind me and the ball was just kind of rolling across the grass. It was a pop fly, and I had totally ignored it. Didn't even see it. I wasn't paying any attention.

"After the game the coach said to my mother, 'Listen, he's obviously not interested, so please don't bring him back.'"

This self-deprecating little story produces another billow of belly laughs at the ten-year-old level. When the softball goes *plop!* in the grass behind the oblivious boy, it's a moment all kids can understand. They've all experienced it in some way. But the idea that when the ball plops in the grass, Tim Samaras simply does not care because he is on an entirely different path, a path that has nothing to do with softball or the expected thing at all, the fact that his alternative path is one filled with excitement, danger, and delight—*that* is a startling and remarkable thought. No doubt some of these kids have been tempted by less traveled paths and new ideas, but it's likely that few have had the courage to really follow them. But now Tim Samaras, chaser of storms, noted scientist, inventor, and fearless explorer, seems to be giving them permission to try.

There's a murmuring in the crowd of kids. Tim is bringing them over to his side, over to his view of the world, over to what he thinks is important and what is not important. The youngsters chuckle and whisper things to each other. Tim seems awake, alive, feasting on this roomful of kid energy. Even though he's been over this material scores of times before—it's his tornado "stump speech"—it's as if he were saying it all for the first time.

He flashes a slide on the screen of all the world's continents, denoted in gray, with tornado activity marked in bright yellow. The center of the United States is a huge incandescent golden blob, brighter by far than anywhere else on the planet.

"There are about 1,300 tornadoes in the U.S. each year, three times more than any other place," Tim says. "We're the tornado kings, man!"

He's really getting worked up now. He starts jumping around, flapping his arms, at one point pretending to be holding one of his "cool orange hats"—the 45-pound HITPR probe he designed to record the barometric pressure inside these most monstrous storms on Earth—over his head to shield himself from imaginary hail. The pantomime produces peals of laughter through the room. He spins around in circles, hands outstretched into the air with the palms shaking like an African dancer, to imitate a giant Doppler radar dish mounted on a truck (which other researchers use to track tornadoes). If enthusiasm is infectious, as it's often said, Tim's youthful audience seems to have caught the bug beyond hope of a cure.

At times he seems to be shooting well over the heads of the youngsters, talking about "what the model data shows," "mobile Doppler radar," and "orders of magnitude." But the kids seem unfazed, and they keep listening intently. Then he flashes a slide on the screen composed of four or five panels of complex meteorological data—something perhaps more suited to a scientific conference than a crowd of ten-year-olds. But the junior scientists just keep listening and watching. After all, Tim might be getting ready to do or say something funny again.

Tim becomes a character in his own story, as in "people say to me, 'Tim, what the heck do you want to take the temperature of a tornado for?' or simply, 'Don't do what Tim does!' " Another character in Tim's story is "crazy Carsten," the bold and impetuous German photographer Carsten Peter, who plays the role of lovable, death-defying stuntman in Tim's narrative. Tim obviously feels a world of affection for Carsten, who keeps popping in and out of the slide show, his red jacket flapping, long hippie hair flying, usually pursued by an enormous tornado looming up behind him like the *T. rex* in *Jurassic Park.* In one lightning sequence, Carsten can faintly be seen standing in the corner of the frame in near darkness, and then in the next frame, taken a millisecond later, still standing there in blinding illumination when a glittering lightning bolt crashes to ground. The kids are spellbound—Carsten is a stand-in for every kid who ever longed to get entirely too close to danger.

At one point Tim shows a slide of himself inside the van he's equipped with electronic gear to operate the ultra-high-speed camera he's developed to photograph lightning. "See," he says, "I'm totally geeked out here." For a moment it's unclear whether he's speaking about himself as a grown man and world-renowned severe-storm researcher or as an eight-year-old boy doing an autopsy on the family blender. Really, it hardly matters; the boy is entirely alive in the man.

At the end of his presentation, he switches to videos he made throughout his storm-chasing career—a kind of "greatest hits" collection of tornado near-death experiences. At one point the video shows him hopping in and out of the chase vehicle, off-loading probes onto the side of a country road as, a quarter mile away, an enormous tornado dwarfs everything around it with its savage dance of wind and dust.

"That's not a good shirt to wear," one kid says, pointing out that Tim has on a T-shirt emblazoned with a circular logo of some sort.

"How come?" Tim says.

"'Cuz it's got a target on it!"

Moments later in the video, Tim's chase partner yells in alarm, "Hurry up, Tim! We don't have enough time! *Seriously!*"

The kids laugh at this, perhaps to release their anxiety at a genuinely alarming moment on the screen.

"Everybody is always telling me to hurry," Tim tells them. "What's up with that?"

The video concludes with a montage of tornado and lightning footage, overlaid with a kind of eerie, keening Celtic music. Chasers keep searching for music that will somehow express the mystery, fear, unworldliness, and dramatic spectacle of tornadoes, and Celtic music seems to come close.

Afterward, kids start peppering Tim with questions:

"Are you going to put big shields over cities?" one kid asks, helpfully, as if Tim were in charge of the world.

Another kid throws out one of those off-the-wall kid ideas: How about if you put your probe inside a boomerang, so it would penetrate the tornado but came back out again?

Other youngsters are surprisingly astute, perhaps because of having watched those cable TV tornado shows on the National Geographic Channel and the Discovery Channel:

"Did you ever chase an F5?" ***"Yes, in Greensburg, Kansas, but it was after dark and too dangerous to get very close."***

"What's the widest tornado you ever chased?" ***"A mile—so wide that it covered the entire town of Manchester, South Dakota, and completely destroyed it."***

"Why did you start doing this in the first place—because you like danger?" ***"Actually, no—I'm not some guy out there trying to get himself killed. I guess I do it for the same reason I took that record player apart when I was a kid: I just want to figure out how it works."***

When the questions die down, Tim wraps it up. His voice changes timbre. He speaks from some deeper place, a place of the soul—a kid reaching out to a roomful of kids.

"I just want to say to you guys that whatever it is that you want to do, whatever you have passion for—that's what you should do. Don't give up. Follow your dreams. You'll never regret it!"

The misfit in the outfield turned his back on softball and went his own way. He followed a path of passion and purpose, becoming one of the premier tornado researchers in the world, contributing to human knowledge about a terrifying atmospheric phenomenon and, perhaps equally important, spending his life doing something he loves. In his self-published memoirs, tornado detective Ted Fujita described a similar sense of joy in a life of deeply fulfilling work: "During my 36 years before retirement, I enjoyed my research much more than anything else. I kept feeling that research is my hobby and hobby is my occupation."

There's a wave of applause and then the kids go swarming up to the front of the room to ask Tim more questions and get their picture taken with him. If there's such a thing as a "tornado celebrity," Tim Samaras is it.

Then the kids go bouncing out of the room, all the future teachers, surgeons, trial lawyers, carpenters, car salesmen, scientists, and perhaps even the severe-storm meteorologist who will one day design a probe that unlocks the forbidding mystery of the tornadic core, laying bare the secrets that have eluded science for centuries.

ACKNOWLEDGMENTS

Tim Samaras

My mom (Marge Samaras) always encouraged me to do my very best—and kept the door to my bedroom closed no matter what, allowing me to have over 25 broken television sets in my room. My dad (Paul Samaras) also encouraged me to do my best, and took the time to help round up all of those television sets when I was 13 years old. I dearly miss you both.

There are also many friends and associates who have contributed to the book and/or my life whom I wish to acknowledge: Larry Brown, supervisor and good friend, gave a 19-year-old kid a chance to do great things. Robert Lynch, fellow engineer and co-worker, has shared lunch with me for 22 years, discussing all of the world's problems, and wrote the software for the HITPR probes. Dean Cosgrove, a good friend and a fellow passionate storm chaser who I chased with in the early years. Brad Carter, a neighbor, friend, and chase partner, who helped deploy my HITPR probes for the first time in 2002 in south-central Kansas. Pat Porter, friend, chase partner, and brother-in-law, has chased with me for 15 years; Pat was with me during the historic Manchester, South Dakota, probe deployment in 2003, documenting everything on video. Todd James, senior editor, photography, at *National Geographic* magazine, always supported my ideas and passions. Good friends Bruce Lee and Cathy Finley joined our team in 2006, collecting valuable data during tornado intercepts; they have also provide excellent scientific review of our collected data and have written several papers of our data collections. Julian Lee (no relation to Bruce Lee), a good friend and co-worker, provided scientific review of probe data and helped conduct wind tunnel testing on the HITPR probes as well as writing several conference papers with me. Al Bedard, senior scientist at NOAA, was the technical monitor of the DOC/NOAA SBIR program that led to the development of the HITPR probe and provided expert counsel and guidance during development; he was also the co-designer of the original TOTO probe in the 1980s. Joe Golden, senior meteorologist at NOAA, was always a proponent of and a source of encouragement for my work with tornadoes. Roy Heyman, an old-school engineer and co-worker, provided the idea for the shape of the original HITPR probe design. Carl Young, a good friend and fellow storm chaser/meteorologist, has been a part of our team since 2003 and always has a positive outlook on life, even if there are no storms. Chris Karstens, Jayson Prentice, and the rest of the Iowa State University students assisted us during the TWISTEX fielding missions the past few years and kindly contributed to some of the pictures in this book. Tony Laubach, meteorologist and good friend, assisted us during our TWISTEX missions and generally kept us laughing most of the time. Partha Sarkar and Bill Gallus, professors at Iowa State University, have collaborated with me on tornado research. Carsten Peter, a free-spirited photographer from Germany, has captured award-winning imagery of our journeys through the plains and has contributed pictures for the book.

My wife, Kathy, has endured countless days as a "chase widow" and has offered continued support for my pursuit of the perfect storm.

Stefan Bechtel

For my beloved little brood, Anya, Sammy, and Milo.

I'd like to express my great thanks to all the people who helped illuminate the dark clouds of my own ignorance about tornadoes and the meteorology of severe storms. First and foremost, I want to thank Tim Samaras, for the privilege of riding shotgun on some of the hair-raisingest rides of my life, and helping me to understand his life's passion in the process. I'd also like to thank Tim's wife, Kathy, for helping me understand Tim. Thanks to meteorologists David Floyd, Bruce Lee, Tim Marshall, Mike Umscheid, and Carl Young for their tornado tutorials. Thanks to storm chasers Matt Biddle, Verne Carlson, Chris Karstens, Chris Collura, Doug Kiesling, Tony Laubach, Jayson Prentice, Kevin Myatt, and David Wolfson, for all the various near-death experiences and stories they shared. David Hoadley was of particular help in telling the story of the early days of storm chasing, and in helping to penetrate the inscrutable "why" behind it all. I did not meet Thomas Grazulis, described to me as the god of tornadoes, but I made great use of his excellent book *The Tornado: Nature's Ultimate Windstorm* in writing this one. The first storm chaser, Roger Jensen, died several years ago, but his spirit and his story touched me. The spirit of Ted Fujita, Mr. Tornado, also hovers over this book, with his glittering insights and his dazzling ability to weave science into art and back again. Thanks to his son, Kazuya Fujita, for recollections of his father's life and work. Thanks to ace photographer Carsten Peter, the Deutschlander, for memorable times in Greensburg, Kansas. I'd like to thank all the former residents of the former Manchester, South Dakota, who shared their memories (some of which were summoned up with tears), especially Rex and Lynette Geyer, Mark Strickler, and Gary Marx.

And a tip of the metaphorical hat to my excellent editors at National Geographic, Barbara Brownell-Grogan and Judith Klein.

SELECTED BIBLIOGRAPHY

"Are You Ready Guide." Federal Emergency Management Agency. Available online at www.fema.gov/areyouready/tornadoes.shtm.

Bluestein, Howard B. *Tornado Alley: Monster Storms of the Great Plains.* Oxford University Press, 1999.

Cerveny, Randy. *Freaks of the Storm: From Flying Cows to Stealing Thunder, the World's Strangest True Weather Stories.* Thunder's Mouth Press, 2006.

Davidson, Keay. *Twister: The Science of Tornadoes and the Making of an Adventure Movie.* Pocket Books, 1996.

Davies, Jon. *Storm Chasers! On the Trail of Twisters.* Farcountry Press, 2007.

Emanuel, Kerry. *Divine Wind: The History and Science of Hurricanes.* Oxford University Press, 2005.

Flannery, Tim. *The Weather Makers: How Man Is Changing the Climate and What It Means for Life on Earth.* Atlantic Monthly Press, 2005.

Fujita, Tetsuya. "A Detailed Analysis of the Fargo Tornadoes of June 20, 1957." Technical Report No. 5, U.S. Weather Bureau Contract Cwb 9530. Department of Meteorology, University of Chicago, 1959.

———. "Memoirs of an Effort to Unlock the Mystery of Severe Storms: During the 50 Years, 1942-1992." Wind Research Laboratory, Department of Geophysical Sciences, University of Chicago, 1992.

Grazulis, Thomas P. *The Tornado: Nature's Ultimate Windstorm.* University of Oklahoma Press, 2001.

Levine, Mark. *F5: Devastation, Survival, and the Most Violent Tornado Outbreak of the Twentieth Century.* Hyperion, 2007.

Mathis, Nancy. *Storm Warning: The Story of a Killer Tornado.* Touchstone, 2007.

Monmonier, Mark. *Air Apparent: How Meteorologists Learned to Map, Predict, and Dramatize Weather*, University of Chicago Press, 1999.

Rankin, William H. *The Man Who Rode the Thunder.* Prentice Hall, 1960.

Stormtrack Technical Library. Available online at www.stormtrack.org/library/.

Wilson, James W., and Roger M. Wakimoto. "The Discovery of the Downburst: T. T. Fujita's Contribution." *Bulletin of the American Meteorological Society* (January 2001), 49-62.

INDEX

Boldface indicates illustrations.

INDEX